THE REPRODUCTION REVOLUTION

A Horizon in Bioethics Series Book from

THE CENTER FOR
BIOETHICS
AND HUMAN DIGNITY

The Horizons in Bioethics Series brings together an array of insightful writers to address important bioethical issues from a forward-looking Christian perspective. The introductory volume, *Bioethics and the Future of Medicine,* covers a broad range of topics and foundational matters. Subsequent volumes focus on a particular set of issues, beginning with the end-of-life theme of *Dignity and Dying* and continuing with the genetics focus of *Genetic Ethics* and the economic and patient-caregiver emphases of *The Changing Face of Health Care.*

The series is a project of The Center for Bioethics and Human Dignity, an international center located just north of Chicago, Illinois, in the United States of America. The Center endeavors to bring Christian perspectives to bear on today's many pressing bioethical challenges. It pursues this task by developing two book series, six audio tape series, six video tape series, numerous conferences in different parts of the world, and a variety of other printed and computer-based resources. Through its membership program, the Center networks and provides resources for people interested in bioethical matters all over the world. Members receive the Center's international journal, *Ethics and Medicine,* the Center's newsletter, *Dignity,* the Center's Update Letters, special World Wide Web access, an Internet News Service and Discussion Forum, and discounts on a wide array of bioethics resources.

For more information on membership in the Center or its various resources, including present or future books in the Horizons in Bioethics Series, contact the Center at:

The Center for Bioethics and Human Dignity
2065 Half Day Road
Bannockburn, IL 60015 USA
Phone: (847) 317-8180
Fax: (847) 317-8153

Information and ordering are also available through the Center's World Wide Web site on the Internet: *www.cbhd.org.*

THE CENTER FOR
BIOETHICS
AND HUMAN DIGNITY

THE
REPRODUCTION
REVOLUTION

A Christian Appraisal of Sexuality,
Reproductive Technologies, and the Family

Edited by

JOHN F. KILNER,
PAIGE C. CUNNINGHAM,
and
W. DAVID HAGER

WILLIAM B. EERDMANS PUBLISHING COMPANY
GRAND RAPIDS, MICHIGAN / CAMBRIDGE, U.K.

© 2000 Wm. B. Eerdmans Publishing Co.
255 Jefferson Ave. S.E., Grand Rapids, Michigan 49503 /
P.O. Box 163, Cambridge CB3 9PU U.K.

Printed in the United States of America

05 04 03 02 01 00   7 6 5 4 3 2 1

Library of Congress Cataloging-in-Publication Data

The reproduction revolution: a Christian appraisal of
sexuality, reproductive technologies, and the family /
edited by John F. Kilner, Paige C. Cunningham, and W. David Hager.
p.   cm. (Horizons in bioethics series)
At head of title: The Center for Bioethics and Human Dignity.
Includes bibliographical references.
ISBN 0-8028-4715-3 (pbk.: alk. paper)
1. Human reproductive technology — Religious aspects — Christianity.
2. Human reproductive technology — Moral and ethical aspects.
3. Sex — Religious aspects — Christianity.   4. Sex — Moral and ethical aspects.
5. Family — Moral and ethical aspects.   6. Family — Religious aspects —
Christianity.   I. Kilner, John Frederic.   II. Cunningham, Paige C.
(Paige Comstock)   III. Hager, W. David (William David), 1946–
IV. Center for Bioethics and Human Dignity.   V. Series.

RG133.5.R465 2000
241′.66 – dc21   99-088803

Contents

INTRODUCTION
THE EXPERIENCE OF REPRODUCTIVE DIFFICULTIES

CONTENTS

PART I
FOUNDATIONAL ISSUES

PART II
SPECIFIC TECHNOLOGIES

Contents

PART IV
RESPONDING TO THE SEXUAL REVOLUTION

PART V
OTHER PROACTIVE RESPONSES

Contributors

Mary B. Adam, M.D., Clinical Lecturer, Department of Pediatrics, University of Arizona College of Medicine, Tucson, AZ, USA

Randy Alcorn, M.A., Director, Eternal Perspective Ministries, Gresham, OR, USA

Nigel M. de S. Cameron, Ph.D., Chairman, The Center for Bioethics and Human Dignity Advisory Board; Consultant, Strategic Futures, Bannockburn, IL, USA

Samuel B. Casey, J.D., Executive Director, Christian Legal Society, Annandale, VA, USA

Susan A. Crockett, M.D., Clinical Professor, University of Texas Health Science Center, Department of Family Practice, San Antonio, TX, USA

Paige C. Cunningham, J.D., Board Chair, Americans United for Life, Chicago, IL, USA

William R. Cutrer, M.D., Director, Gheens Center for Christian Family Ministry, The Southern Baptist Theological Seminary, Louisville, KY, USA

Joseph L. DeCook, M.D., Physician, private practice, Holland, MI, USA

Thomas E. Elkins, M.D., Chief of Gynecology and Visiting Professor, Johns Hopkins School of Medicine, Baltimore, MD, USA

CONTRIBUTORS

Robert W. Evans, Ph.D., President, Veritas Ministries, Auburn, CA, USA

Jeannie W. French, M.P.H., Executive Director, National Women's Coalition for Life, Pittsburgh, PA, USA

W. David Hager, M.D., Physician, Central Baptist Hospital and University of Kentucky Medical Center, Lexington, KY, USA

Donna Harrison, M.D., Physician, Lakeland Regional Health Systems, Berrien County, MI, USA

Camilla Hersh, M.D., Clinical Assistant Professor, Department of OB/Gyn, Georgetown University Medical School, Washington, DC, USA

Dennis P. Hollinger, Ph.D., Dean of College Ministries and Professor of Christian Ethics, Messiah College, Grantham, PA, USA

C. Christopher Hook, M.D., Hematologist/Ethicist, Mayo Clinic, Rochester, MN, USA

Teresa Iglesias, D.Phil., Lecturer in Philosophy and Medical Ethics, National University of Ireland, Dublin, IRELAND

Charlene Q. Kalebic, J.D., Law Clerk to the Honorable Morton Denlow, United States District Court for the Northern District of IL, Chicago, IL, USA

John F. Kilner, Ph.D., Director, The Center for Bioethics and Human Dignity, Bannockburn, IL, USA

Walter L. Larimore, M.D., Physician, Heritage Family Physicians, Kissimmee, FL, USA

Joe S. McIlhaney Jr., M.D., President, The Medical Institute for Sexual Health, Austin, TX, USA

Gilbert Meilaender, Ph.D., Board of Directors Chair in Christian Ethics and Professor of Theology, Valparaiso University, Valparaiso, IN, USA

C. Ben Mitchell, Ph.D., Assistant Professor of Bioethics & Contemporary Culture, Trinity International University, Deerfield, IL, USA

Dónal P. O'Mathúna, Professor of Bioethics and Chemistry, Mount Carmel College of Nursing, Columbus, OH, USA

Contributors

R. Martin Palmer, J.D., Attorney at Law, Hagerstown, MD, USA

Scott B. Rae, Ph.D., Professor of Biblical Studies and Christian Ethics, Talbot School of Theology, Biola University, La Mirada, CA, USA

Charles M. Sell, Th.D., Professor of Educational Ministries, Trinity International University, Deerfield, IL, USA

Joyce A. Shelton, Ph.D., Associate Professor of Biology, Trinity International University, Deerfield, IL, USA

Gracie Hsu Yu, M.H.S., Consultant, Family Research Council, Washington, DC, USA

Joleen Zivnuska, M.S.N., Women's Health Nurse Practitioner, The Center for Reproductive Medicine, Wichita, KS, USA

Preface

Few social or technological developments in history have captivated people's imagination or raised more ethical questions than today's reproduction revolution. In this book we have gathered together a stellar team to assess the impact of this revolution on sexuality, childbearing, and the family.

There are certain foundational issues that must be addressed by anyone interested in dealing with these matters. For example, does having the technological capability to carry out various scientific endeavors give us the moral right to pursue those challenges? Does the fact that an act is legal make it morally right? Do worthy outcomes (babies) justify any measures necessary to obtain them? Is procreation an innate right or a God-given blessing to be received responsibly?

At the same time we must take seriously the consequences of behavior. All too often we insist on doing whatever we can, by whatever means; and if there are adverse consequences, we expect to be absolved from them. But we cannot escape so easily. Data from youth risk behavior studies indicate the tragic results of poor choices in sexuality and other potentially addictive behaviors. Moreover, youth are not alone in making such inadvisable choices. This book challenges us all to be more ethically responsible — especially for the sake of our children — for the sake of *all* children.

How important is it for children to be the fruit of a one-flesh love relationship rather than merely of donors to the cause? Is caring for the orphaned, abandoned, and misdirected children of the world as much our responsibility as developing technology to allow the infertile to reproduce? Is a human embryo truly a child — simply at an early stage of growth — or is a human embryo, ultimately, discardable research material (regardless of the "respect" we may profess for "it")? Is technology actually a gift which can better enable us to do that which God calls and directs us to do in this life, or is it a vehicle to facilitate whatever we wish to pursue? Such questions — at once both profoundly human and yet unavoidably theological — are at the heart of this book.

An adequate analysis of these matters, however, must include legal analysis as well. The law tells us something about ourselves. Sometimes it reflects the trends and pressures of changing technologies, values, and attitudes. Sometimes it tells us what we are willing to give up in order to protect what we want to keep. Sometimes it dictates to us through the voice of a Supreme Court — telling us what is legal, democratic, or constitutional. That Court announces constitutional rights and fundamental freedoms with the intent of settling deep questions once and for all. But the most troubling and hotly contested moral issues cannot be "settled" so easily.

Whether the question is cloning, abortion, surrogate motherhood, the disposal of embryos created by in vitro fertilization, or assisted suicide, our legal institutions have failed to grapple adequately with the essential problem: What does it mean, under law, to be a human being, a member of the species homo sapiens, regardless of stage of dependency or biological development? Can ethics and the law draw any line of protection around the most vulnerable among us? That is a question of great concern to Christians — a question Christians must raise in the larger society as well. The present volume demonstrates that a Christian perspective has much to offer to a broad audience in terms of wise counsel.

The book is organized into an Introduction and five major sections. The Introduction and first major section provide an important backdrop to all that follows. The Introduction orients the reader to the experience of reproductive difficulties through the lives of a patient (Jeannie French), a nurse (Joleen Zivnuska), a medical educator (Thomas Elkins), and a physician (William Cutrer). The first major section examines four important foundational issues. Nigel Cameron begins by exploring the ambiguous triumphs of technology that have succeeded in separating sex and reproduction. Gilbert Meilaender then evaluates the ethical price that people are willing to pay in order to produce a child of their own. In the next chapter, Dónal O'Mathúna demonstrates how a clear understanding of the goals of medicine can provide a critical perspective from which to assess sexual health drugs such as Viagra. Robert Evans concludes the section with a discussion of the moral status of embryos.

Building on such foundations, the authors of the second major section offer an ethical analysis of specific reproductive technologies. Dennis Hollinger opens the section with a framework for sexual ethics that provides ethical parameters within which reproductive technologies such as intrauterine insemination and in vitro fertilization should operate. Teresa Iglesias more specifically focuses on the use of donor eggs and sperm in such technologies — seeing there the tragic face of anonymity. In the following chapter, Scott Rae provides an ethical analysis of the age-old yet technologically rejuvenated practice of surrogate motherhood. John Kilner concludes the section with an assessment of the imminent practice of human cloning.

With all of this analysis as a frame of reference, the third major section takes up two particularly difficult cases. The first is a legal case involving a surrogate mother who wants to abort the fetus she is carrying despite the vehement opposition of the woman who has contracted her to bear her child. Following an introduction to the case, plus the text of the case itself, Christopher Hook and Ben Mitchell analyze the array of issues the case raises — especially the intricacies of the surrogacy arrangement. Martin Palmer and Joyce Shelton then consider often-overlooked aspects of the proposed abortion. Samuel Casey closes the discussion with an examination of the ethical and, especially, legal issues involved in the embryo freezing that this case entails. The other difficult case — whether or not using the birth control pill causes abortions — is next addressed in the form of a debate. Following an introduction by special debate editor Linda Bevington, authors Randy Alcorn and Walter Larimore argue that the available scientific data appear to indicate that the Pill probably is indeed abortifacient and therefore its use is unethical. Joseph DeCook, Donna Harrison, Susan Crockett, and Camilla Hersh counter that the scientific data are inconclusive and so decisions about the Pill's use are rightly left to one's biblically informed conscience.

In the fourth major section, the book broadens its scope by investigating the sexual revolution that has accompanied the emergence of new reproductive options. David Hager opens the section by discussing youth risk takers, whom he identifies among the casualties of the sexual revolution. In a similar vein, Joe McIlhaney examines historical and current sexual practice, with an eye to implications for the future. Mary Adam then explicitly considers the potential of preventive public policy for curbing sexually transmitted disease in today's postmodern world.

The final major section carries this proactive orientation a step further by presenting three constructive responses needed to give positive direction to the reproduction revolution currently underway. Gracie Hsu Yu begins by examining the twin strategies of making laws and changing hearts. Focusing on the legal strategy in greater detail, Charlene Kalebic demonstrates how the law can proactively address a particularly controversial upcoming technology like human cloning. Finally, Charles Sell closes the book by explaining why, in the end, the most important strategy for participating in the reproduction revolution may simply be to affirm the importance of the family.

The word "revolution" is not used lightly here. The very term "reproduction" embodies a fundamental shift from the implicitly God-honoring term of "procreation" to the human-centered manufacturing language of reproduction. The tools of that manufacturing have become as lethal as the weapons in any military revolution. Yet, revolutionary times can also be extraordinarily renewing times, in which people resolve to continue no longer in certain oppressive and destructive ways. May it be so in our day.

John F. Kilner, Ph.D.
Paige C. Cunningham, J.D.
W. David Hager, M.D.

INTRODUCTION

THE EXPERIENCE OF
REPRODUCTIVE DIFFICULTIES

A Patient's Perspective

Jeannie Wallace French, M.P.H.

As technological advances occur in the area of human reproduction, would-be parents are faced with decisions that people have never before had to make. Not only are new options for increasing fertility available, but once conception occurs, parents must make treatment decisions for unborn children whose life-threatening or disabling conditions can now be identified. I offer my own experience in this arena in the hope that it may help clarify some of the ethical challenges before us so that we may wrestle with them more effectively.

I am a thirty-seven-year-old health care professional with a master's degree in public health. I am a diplomate of the American College of Health Care Executives, and a member of the Chicago Health Executives Forum. This story, however, is about being a mother.

In early 1992, after two years of difficulty in conceiving followed by an early miscarriage, my husband Paul and I sought fertility counseling. One course of ovulation-stimulating medication (Chlomid®) resulted in a pregnancy that ended in fetal death at eleven weeks.

In March of 1993, again with fertility support (Pergonal®), I conceived triplets. At nine weeks, spotting was an early sign that we had lost one of our unborn babies. By early summer, the twin pregnancy was confirmed, and my husband and I joyfully celebrated making it through the first trimester. We were thrilled to have a multiple pregnancy after so many years of failed attempts to become parents.

Our joy was short-lived. At four months, doctors at Rush Presbyterian–St. Luke's Medical Center in Chicago noticed that one twin, our daughter Mary, showed growth retardation most likely caused by an abnormality.

3

Through a series of high definition sonograms over the next several weeks, Mary was diagnosed with an occipital encephalocele — a condition in which the majority of the brain develops outside of the skull.

We were devastated. The chief radiologist advised us that the chance of Mary surviving even the pregnancy was slim. We learned that if she were born alive and were strong enough, she would need immediate, extensive life-saving surgery to remove the encephalocele and to reconstruct her skull.

At the first communication of the diagnosis, I headed to the Loyola University Medical School library, located only a mile from our home. I was determined to find someone, somewhere, that could change the prognosis for our daughter. Instead, I was horrified at the pictures I saw of the babies who had been afflicted with this condition. Every medical source I read advised termination of the pregnancy. In our case, we were advised to abort Mary while leaving her twin to continue developing.

Although all of the medical textbooks encouraged abortion for our daughter, a friend eventually pointed us in the direction of a handful of neurosurgeons who were doing corrective surgery on children like Mary. They were eager to help correct the disfiguration but honest about the limited capabilities our daughter was likely to have. Even if Mary survived the surgery, she would lose the part of her brain that had developed outside the skull. Mary had virtually no possibility for normal development. She would likely never walk, talk, or learn.

The radiologist at Rush Medical Center wanted to do an amniocentesis to further confirm the diagnosis. When we told him that we would not abort our daughter regardless of her condition, the doctor suggested that there was "no point" in doing the test. It was then that we were advised that if Mary died in utero, I might go into labor and we would likely lose the other twin.

Medical specialists, articles, and textbooks alike recommended abortion for Mary due to her severe disability and to reduce the complications of the twin pregnancy. But my husband Paul and I both felt in our hearts that our baby girl was already a beloved member of our family, regardless of how "imperfect" she might be. We felt that she was entitled to her God-given right to live her life, however short or difficult it might be, and if she were to leave this life, we believed she had a right to die peacefully.

After confirming the severity of baby Mary's condition, I began to research organizations that offered support to families with special children. I found countless parents who had lived through similar experiences and were willing to help me explore options that would enrich Mary's life if she survived. I even located a little boy, also in Chicago, who had the same condition. He was three years old and functioning as an infant.

And while we planned to raise Mary in a loving home with community and family support, I also was surprised to find organizations of would-be adoptive parents, just waiting for children with special needs. The National Down Syndrome Adoption Exchange lists 100 to 150 couples, all hoping to bring a child with Down Syndrome into their family. There is a similar waiting list for spina bifida babies. Adoptive parents are also waiting for terminally ill babies, including those with AIDS.

Near the end of the pregnancy, we learned that Mary's twin was a seemingly healthy baby boy, who we would later name Will. How amazing it was to hear of the difference in treatment suggested for the two children that were growing side by side within me. No one ever suggested aborting Will in order to focus more completely on treating our daughter. But the medical community was clear about Mary. A disabled baby was not an outcome they supported. There was a constant pressure to "take care of things" without any help in handling the special needs of our sick infant.

Since we were not interested in aborting our daughter, our obstetrician helped me make her as comfortable as possible. I spent the last few months of my pregnancy on bed rest lying on my left side to ensure optimal oxygen transport to the twins. Our only health complication was my development of a slight heart murmur. Further examination eventually ruled out any serious problem.

We learned that baby Mary, like anencephalic babies, would not survive the uterine contractions of normal vaginal delivery. We petitioned for a cesarean delivery in order to spare our daughter what we thought might be excruciating pain. A cesarean delivery appeared to be our only chance to see Mary alive.

Baby Mary Bernadette French and her healthy brother William David were born a minute apart on the morning of December 13, 1993. Little Will let out a hearty cry and was moved to the nursery. Our quiet little Mary remained with us, cradled in her father's arms.

In light of the severity of Mary's condition, doctors expected her to die immediately. But our little girl was a fighter. She survived to meet her grandparents and was held, loved and sung to for six hours. At 3:30 P.M. on the day she was born, baby Mary peacefully died.

Mary's calm birth and death do not end her story, however. Three days after Mary died, on the day she was laid to rest at the cemetery, my husband Paul and I were notified that her heart valves were a match for two Chicago infants in critical condition. Baby Mary's ability to help save the lives of two other children suddenly proved to many others what Paul and I already knew on other grounds — that her life had value.

We had learned through the death of an infant in our neighborhood that organ donations among infants were rare. In fact, our conversations with the Regional Organ Bank of Illinois revealed that parents are usually too shocked at the loss of an infant to consider donations. Additionally, those parents who know they carry a severely disabled child almost always abort these babies, eliminating the opportunity to donate.

Our daughter Mary's life lasted 37 weeks, 3 days, and 6 hours. In effect, like a small percentage of children conceived in our country every year, she was born dying. We did no more than any other parents would for a dying child: we gave our baby love and comfort. We did not need to help her die.

Throughout the pregnancy and delivery, Paul and I felt all the normal feelings of beginning our family — joy, fear, anxiety, and overwhelming responsibility. But this pregnancy put us at odds with the medical experts in our community. Aside from my obstetrician who knew our values, there was no support for bringing our disabled baby to term; no one offered to connect us with other parents, no one provided a helpful list of specialists who would work with us. It was our own commitment and determination that enabled us to uncover what we did.

Paul and I could not save our little girl, but we did our best to give her a good life for as long as she lived. At the delivery, we sent our healthy boy to the nursery right away, feeling we would have a lifetime to love, hold, and marvel over him. We welcomed our sick little baby girl with open arms, not knowing how long our time together would be.

I don't know what kind of a mother I would have been to my daughter Mary, had she survived. My heart goes out to parents faced with the reality of raising children who are physically or mentally challenged. Throughout my pregnancy, I felt their sadness and faced their fears. I somehow believed that we would find the strength and support to love and raise her, like so many parents of special children have done throughout history.

In these years following Mary's short life, I have been blessed to provide one-on-one counseling to parents who have received bad news about an unborn child's medical condition. I have become even more committed to the consistent ethic of life, valuing each individual as created in the image of God, regardless of the physical afflictions of a fallen world.

I have also listened to brokenhearted women who weep after the abortion of their disabled babies. I have witnessed firsthand the grief that parents experience when they abort a child believed to be sick. And I can only imagine the immeasurable pain felt by these disabled children when they are aborted.

When a poor prognosis regarding an unborn child is communicated, parents are swept into overwhelming emotion. Unable to visualize the beautiful, perfect baby we have hoped for, we panic. We want to love our child, but are thwarted by the described disfiguration of our baby. For severely disabled children, like little Mary, the medical community responds by urging parents to "take care of the problem" through abortion. And unwittingly, parents accept abortion as the humane end for these sick children. It is a deeply regrettable act, void of true compassion.

Unlike others I have comforted after abortion, Paul and I have no regrets about letting our daughter be born. We do not lose sleep over pain she might have suffered had we opted to end her life. We felt confident that we simply acted as parents, letting God reclaim this little life peacefully. Needless to say, there were also other reproductive decisions that led up to our final decision. These, too, must be subjected to the kind of ethical scrutiny undertaken in the present volume.

It is a fact of life that some children are born dying. If we are to call ourselves a civilized culture, we must support parents in saying good-bye to them peacefully. And if other children face a life of disability, let us welcome them into this society, with our arms open in love. Who could possibly need us more?

A Nurse's Perspective

Joleen Zivnuska, M.S.N.

I work in a reproductive endocrinology and infertility practice of fertility specialists who perform assisted reproductive technology (ART) procedures such as in vitro fertilization (IVF) and gamete intra-fallopian tube transfer (GIFT). As a patient with a diagnosis of severe endometriosis, I first became acquainted with the medical field of infertility twelve years ago. I have undergone multiple surgeries to aid in pain relief as well as increase my chances for a pregnancy. Severe endometriosis results in a decreased ability of the ovaries to produce multiple follicles.[1] It is also associated with decreased fertility rates. Anticipating that we would not be faced with decisions regarding the deposition of a large number of embryos helped us feel comfortable in our decision to pursue treatment via ART procedures. However, both IVF and GIFT procedures were unsuccessful in my case. Thankfully, we have been able to adopt two children. Having personally faced the ethical decisions presented by ART, I have experienced feelings and frustrations similar to those of many of my patients.

A few statistics may be helpful here in terms of background.[2] Infertility is defined as the inability to achieve a pregnancy after one year of unprotected intercourse (after six months if the woman is over age 35) or the inability to carry a pregnancy to live birth. Evaluation of the infertile couple must focus on both partners, as 40 percent of infertility is due to a male factor, 40 percent to a female factor, and 15 percent to both; the cause of the remaining 5 percent is unexplained. To evaluate only one partner, even when the problem

---

1. Leon Speroff, Robert Glass, and Nathan Kase, *Clinical Gynecologic Endocrinology and Infertility,* 5th ed. (Baltimore: Williams and Wilkins, 1994), p. 940.
2. Speroff, Glass, and Kase, *Clinical Gynecologic Endocrinology,* p. 809.

seems apparent, is inappropriate due to the high incidence of multiple problems. All too often a couple is referred to us after the wife has had a complete infertility evaluation including a laparoscopy but no one has evaluated her husband with a semen analysis.

Infertility affects one out of every six couples in the United States.[3] Primary infertility occurs in couples who have never before achieved a pregnancy. Secondary infertility occurs after one partner has previously achieved a pregnancy. These couples may not receive as much empathy from friends; however, they are equally frustrated, as they know they have been able to conceive and give birth previously. They often feel defeated in their inability to control their bodies and produce this greatly wanted second or third child. Their pursuit of pregnancy can be even more zealous than that of primary infertility couples, for they already know that parenting is a wonderful experience. I often validate their frustration, as some feel a sense of false guilt knowing they already have one child at home.

Infertility is a crisis for many couples. It can affect them in all areas of their lives: emotionally, relationally, vocationally, physically, and spiritually. It may represent the first obstacle that the individual or couple has not been able to overcome in fulfilling their lifelong dreams. They can feel vulnerable, helpless, and even angry. In fact, infertility involves many losses that need to be identified, validated, and grieved by each couple. Pat Johnston, in her book *Taking Charge of Infertility*,[4] describes six losses many couples experience due to their infertility: (1) control over many aspects of life, (2) individual genetic continuity linking past and future, (3) the joint conception of a child with one's partner, (4) the physical satisfaction of pregnancy and birth, (5) the emotional gratification of pregnancy and birth, and (6) the opportunity to parent. These are intangible losses, often totally unrecognized by family and friends. If a stillbirth occurs there is a funeral, flowers are sent, and family and friends join the couple to grieve their loss. Miscarriages or negative pregnancy tests are often losses acknowledged and grieved by the couple alone.

Family and friends unfamiliar with infertility losses often make very hurtful comments in their ignorance. A list of helpful resources, including suggestions for what not to say and what people can do to help couples experiencing infertility, is included in an appendix at the end of this chapter. I encourage couples to copy this material and send it to family and friends to help educate them and elicit helpful support. Resolve, a national organization, of-

3. Speroff, Glass, and Kase, *Clinical Gynecologic Endocrinology*, p. 809.
4. Patricia Irwin Johnston, *Taking Charge of Infertility* (Indianapolis: Perspective Press, 1994).

9

fers information, advocacy, and support services to infertile couples.[5] Bethany Christian Services offers adoption, family social services, and an infertility newsletter entitled "Stepping Stones."[6] The current editors of "Stepping Stones," John and Sylvia Van Regenmorter, have also co-authored one of the best Christian books on infertility, *Dear God, Why Can't We Have a Baby? A Guide for the Infertile Couple.*[7]

The office staff may be the only other people who know of the couple's infertility struggle. Nurses and the office receptionist often receive the brunt of a patient's frustration and anger. Patients often identify physicians as the enablers of their dreams, therefore, they do not want to risk alienating them with their anger. Thankfully, nurses are also one of the first to receive hugs from patients.

The fast pace, demands for flawless work, and ever-changing protocols turn most of the nursing staff into obsessive-compulsive employees. High expectations plus working with very motivated, sometimes angry, and often demanding patients contribute to the burnout of reproductive endocrinology nurses. The majority of nurses working in this field make a career change after four to five years.[8] A California colleague shared a story of how motivated a couple can be. One couple navigated their way to the clinic on time for their IVF egg retrieval the morning after the devastating earthquake in Los Angeles several years ago. Staff members were taking care of their families and what was left of their homes; yet obviously this couple expected that everyone would be at the clinic to complete their egg retrieval.

My role as a nurse practitioner also includes being a patient advocate and counselor. Approximately 60 percent of my time is spent attending to the emo-

5. Resolve
   1310 Broadway, Somerville MA 02144
   Phone: 617-623-0744
   Web site: *www.resolve.org*
A national nonprofit network of state chapters offering information, referral, support, and advocacy services to infertile couples.

6. Bethany Christian Services
   901 Eastern N.E., P.O. Box 294, Grand Rapids, MI 49501-0294
   Phone: 616-224-7617
   Web site: *www.bethany.org/stepping*
An international not-for-profit, pro-life, pro-family organization. Bethany offers adoption and family social services. "Stepping Stones" is an infertility newsletter published bi-monthly.

7. John and Sylvia Van Regenmorter, *Dear God, Why Can't We Have a Baby? A Guide for the Infertile Couple* (Grand Rapids: Baker Book House, 1986).

8. Judith P. Bernstein, Ph.D., Department of Maternal and Child Health, Boston University School of Public Health, Boston. Unpublished data, 1995.

tional aspects of infertility. Thankfully, I work in an office that values that time, so my appointments are scheduled thirty minutes apart. The exam may take five minutes; the rest of the time is an opportunity to reassess the treatment plan and offer emotional support. I try to encourage each couple to take inventory and decide how much money, time, effort, physical discomfort, and emotional distress they are willing to endure in their quest for a pregnancy. This enables them to develop *their* treatment plan and restore some of the control in *their* lives.[9] My aim, when appropriate, is to leave each patient encounter with a hug. I want patients to know I also care about them on an emotional level.

As mentioned previously, infertility often leaves couples feeling out of control of their lives. They are unable to make their bodies produce eggs or sperm to fulfill a lifelong dream. Nurses and physicians also feel the frustration of not being able to make a pregnancy happen. We are humbled every day with the fact of who is really in charge of every conception.

I strive to empower couples to grieve their losses and make decisions regarding their treatment plan so that they can grow as a couple and reach a higher plane in their relationship. Many couples have grown so much closer during their infertility journey that their decision to pursue child-free living becomes a solace. I am as excited to see pictures of a new sailboat from these couples as I am a baby picture from others.

One of my roles is to coordinate our male factor infertility program, including intrauterine insemination. I often recognize gender differences in grieving and coping with infertility. Women tend to be verbal, sifting through decisions as they discuss them, whereas men often process these decisions internally, speaking only when decisions have been made. I review these differences in my consultations with the couples, hoping to avoid or lessen future misunderstandings due to their unique coping styles.

As a patient advocate, I am very concerned about a deceitful infertility center misleading vulnerable patients. While an allegedly unethical fertility center does occasionally make headlines, stringent regulations are in place in countries such as the United States. The American Society for Reproductive Medicine (ASRM) formed the Society for Assisted Reproductive Technology (SART) as a watchdog organization.[10] To maintain SART membership, em-

9. Pat Johnston describes this whole process in greater detail in her book *Taking Charge of Infertility.*

10. American Society for Reproductive Medicine
    1209 Montgomery Highway
    Birmingham, AL 35216-2809
    Phone: 205-978-5000
A national organization for infertility specialists and interested professionals. It provides

bryology laboratories (where ART procedures are completed) are regulated not only by state health agencies but also by the Department of Health and Human Services and the Health Care Finance Administration (HCFA) under the Clinical Laboratory Improvement Act of 1988 (CLIA 88). The Federal Trade Commission (FTC) regulates ART advertising. Annual reporting of success rates is required by the Fertility Clinic Success Rate and Certification Act of 1992 (Wyden Bill); this was published for the first time on the Internet in December 1997 by the joint efforts of the Centers for Disease Control (CDC), Resolve, and SART/ASRM. Since 1989, SART programs have voluntarily reported their clinic pregnancy rates. SART members must agree to on-site validation inspections of their data as reported to the CDC. Couples are encouraged to pursue ART procedures from a center with current SART membership. SART reports a pregnancy only after a gestational sac has been visualized via ultrasound. These regulations not only dictate quality control but also provide the forum for all centers to uniformly compile and report their pregnancy outcome data.

Each fertility center decides what procedures it will provide. The center where I work has made some restrictions in its IVF and frozen embryo program. We impose a limit on the number of embryos to be returned to the uterus, depending on the woman's age and embryo quality, as we are very concerned about limiting the possibility of a multiple pregnancy. For couples who are uncomfortable freezing embryos there is the option of limiting the number of eggs to be fertilized. Non-ART treatment protocols are quite common due to their lesser cost. It is imperative to monitor patients with ultrasound to determine the number and maturity of the ovarian follicles being produced (only mature follicles release eggs capable of being fertilized). We will cancel the cycle and instruct the couple not to have intercourse if too many mature follicles are present. Furthermore, we have chosen not to participate in surrogacy arrangements.

The multitude of ethical issues that I confront daily challenges both my mind and soul. Christ has called his followers to minister to the hurting, to bind up the wounds of the brokenhearted. He has called me to serve on the margin — to be a Christian voice regarding the ethics in our office. He has called me to be a voice of faith, an instrument of peace, to speak into our patients' often confused, lonely, vulnerable, and hurting lives. Admittedly, patients and health care professionals who are Christians of good faith sometimes hold different views

---

medical referrals by listing member physicians in one's area but does not give recommendations. It also publishes the medical journal *Fertility and Sterility* and offers continuing medical education conferences.

on which treatment options are ethically acceptable. This fact only underscores the need for us to become educated about and wrestle with the ethics of these options. Our goal should be to serve infertile believers and unbelievers alike with the same sensitivity and compassion bestowed upon those who have experienced the physical death of a child.

Appendix

Renee Cristiano, LSCSW, is a clinical social worker who sees many of our patients. She has put together these helpful suggestions for those unfamiliar with the experience of infertility.

Things Not to Say

1. "Sex must be fun" or "look at all the practice you get!" Passion may fly out the window when intercourse must be scheduled. The scheduling of sex can be embarrassing and frustrating; the couple may feel pressure.
2. "Just relax." This statement minimizes a very complicated problem. It may induce guilt or the feeling "I must be causing this."
3. "I know someone who adopted and then got pregnant." Although you may know someone to whom this happened, it occurs very infrequently. Adoption is a decision that deals with choice, a path toward parenthood. It is not a method that resolves infertility.
4. "Why don't you try harder?" These people are already trying as hard as they can and that is the frustration. They do not have control; they do not have the power to make the pregnancy happen.
5. "Be glad you don't have kids" or "here, take mine." You are not serious about giving away your children, and it belittles their pain.
6. "I know what you are going through." Unless you have dealt with infertility personally, you do not know.
7. "Why don't you try . . ." Do not offer suggestions for treatment. They are following what may be complicated directions for treatment. They need support, not distractions.
8. "It's God's will." For some, infertility does raise spiritual issues. It is a personal issue. They need support, not judgment. What may sound comforting to you may sound to them as if God is punishing them.
9. (If a pregnancy did occur and was lost) "See, you can do it, try again." They may feel grief and need to be allowed to express it.

10. "You don't have kids, you must have lots of money." For many couples, infertility treatment and evaluation is costly, and their insurance may pay little or none of the costs. They may have made great personal sacrifices to try to resolve the problem.

What You Can Do

1. Ask for direction: "How can I help you? What do you need from me now?" The person may not be able to say what he or she needs, but will feel better for your having asked and not ignored them or presumed.
2. Listen without judging or offering advice.
3. Use feeling words and reflection. For example: "It seems you're telling me you are upset," "Are you feeling sad about this?"
4. Ask "Do you want to talk about this or be quiet?" They may have difficulty setting a limit, so you may have to read subtle cues as to whether or not they want to talk.
5. Say "I'm sorry you hurt."
6. Do not ask for details regarding treatment; respect their need for privacy. Say "You can tell me what you need me to know."
7. Offer help and support. For example, offer to go with the person when procedures are necessary, cook a meal, send a note. All of these send strong messages about concern and care while respecting the rights of the person.
8. Holidays can be difficult, especially Christmas. Total focus of families on children can be very painful and difficult to handle. Be sensitive and understanding if the person feels that he/she needs to withdraw from some of the activities. Have some adult time if possible.

A Medical Educator's Perspective

Thomas E. Elkins, M.D.

What is ideal reproduction? A male and a female, both adults, enter into a marriage covenant with a lifelong commitment to love one another. From this relationship grows a love that results in a glorious spiritual, mental, and physical union that is not unlike the creative, dynamic love that God has for us. From this loving sexual union, a genetically, physically, and mentally new life joins this divinely created world in which we all share as those produced in the image and likeness of our creator.

In this ideal reproductive situation, there are no miscarriages or abortions, no maternal deaths, no infections following delivery, no ectopic pregnancies, no reproductive diseases, no infertility, no birth control, and no new reproductive technologies. Sadly, the realities of our fallen world are otherwise. I have spent a significant amount of time in Africa and have seen the incredible suffering of the developing world in terms of what women go through in reproductive health. The situation there does not compare to the situation in the United States. In rural West Africa today, 1 in every 12-14 women will die in childbirth. There are as many women dying in pregnancies and childbirth every week in Africa as there would be if two 747s full of pregnant women collided in midair and everyone on board were killed. In fact, the true figure is probably much higher. From 40 to 70 women die for every 1,000 deliveries in rural West Africa in the remote areas and towns. In the major cities, major universities, and big hospitals in West Africa, between 10 and 14 women die for every 1,000 deliveries. Comparing that to 1 in every 10,000 deliveries in the United States, one begins to understand the significant difference. One begins to see how far technology has brought the more developed nations of the world.

Technology is not the enemy. It is the use of it in unethical, God-dishonoring ways that is the problem. There is no substitute for careful, prayerful discernment. Caution is needed lest in our zeal to remove the weeds we lose the benefit of the nourishing, life-sustaining plants as well.

Discerning the acceptable uses of reproductive technologies is especially difficult because the "natural" alternative in our fallen world is so destructive of embryonic and fetal life. Studies estimate that from 25 to 65 percent of pregnancies end up in miscarriages. That is a high number of conceptuses that are either not implanting or are aborting usually very early in the pregnancy. Once we decide to try to have a child — a morally acceptable action for a married couple — we have decided to put embryos at significant risk. Such risk is not created only by the use of reproductive technologies — it is an unavoidable but necessary risk of all childbearing.

I have referred here to childbearing as "a morally acceptable option" for a married woman. Actually, it is more important than that: Infertility is something we must adequately understand if we are to assess the benefits and burdens of reproductive technologies accurately. In the United States, we view infertility as a sad situation, as a tragic situation. Yet, it seems that Americans have so many blessings in life. They can all get by. In this light, infertility is seen as a minimal tragedy compared to so many others. We can adopt; we can open our homes and our lives to children in many ways.

In West Africa, however, infertility for many is not only a social concern; it is a life-and-death concern. In rural Ghana and Nigeria, where I have worked with so many of the different tribes, to be infertile is essentially the end of a woman's life as she has known it. In some of the most remote villages it is not unheard of to poison infertile women simply because there are too many mouths to feed and these women are not of use in the society. Alternatively, they may be almost enslaved to a life of labor since they are unable to fulfill their main purpose. When we begin to understand what infertility means worldwide, we begin to sense the tragedy of infertility that women feel even in more developed nations.

Life is a precious gift of God that we must energetically protect and support. That means we should not engage in childbearing in a way that subjects embryos to significantly greater risk than is unavoidable; and we should do all that we can to improve women's lives within this same ethical guideline. One implication of this outlook is that people within a marriage setting can, with informed consent and within limits, engage in some new reproductive technologies. They should do all they can do to ensure that every embryo created is implanted, while limiting the number created so that pregnancy will not be dangerous for the mother or the developing children.

Why limit the use of reproductive technologies to married people? Children need two-parent families — male and female two-parent families. It is just that straightforward. The five years I spent in New Orleans, where I was chairman of Ob-Gyn at Louisiana State University and then head of their reconstructive pelvic surgery program, was an eye-opening time in one of the classic inner cities of America. Fatherless homes are the rule there. Growing up in such an environment often leaves young people with an inability to see a way into the future, and leads some to turn to crime. Young women also face terrible economic problems as they seek to raise children without the support of the children's fathers.

To think that we are now intentionally bringing children into the world without fathers through the use of reproductive technologies absolutely baffles me. Our inner cities cry out for society to use some common sense and not to use the technological advances that we have been blessed with to create the very family situations that we should be striving to avoid. This is not a problem solved with money. Even without the economic difficulties, intentionally bringing children into the world without the fathers from whose sperm they were created deprives them of one of their parental anchors in life, even if someone steps in to play a parental role. The children know that one of the parents who genetically gave them life did not want to keep them — a traumatic realization.

The challenge today is to care deeply for *both* parents and children. We must do all we can to relieve the pain of such tragedies as infertility. But we must make sure that we do not intentionally deprive children of their genetic parents in the process.

A Physician's Perspective

William R. Cutrer, M.D.

I have a wonderful opportunity to teach a four-week series on medical ethics to the Ob-Gyn residents at Baylor Hospital in downtown Dallas. I am a graduate of this institution and practiced there for fourteen years following my residency. When I was in training, no ethics course existed nor did annual requirements for continuing medical education in ethics. The residents and I discuss the historical ethical views, various systems and principles, learning the vocabulary and paradigms critical to meaningful participation in the ethical arena.

In one of our meetings this past year, a third-year resident presented for discussion the following actual case (reality provides all the best material):

> A young married woman presented at term in labor with her first child. This would not seem particularly unusual except for the interesting collection of "family" in the labor room. The father of the child was actually the patient's stepfather, and she was carrying a pregnancy conceived through intrauterine insemination because her biologic mother had been surgically sterilized. Thus the daughter was a "traditional surrogate" for her natural mother. The patient was about to deliver her own brother or sister!

The residents, after some audible moans and groans, pondered aloud, "Is this incest? Is it adultery? Is it wise?" I asked them: "What do you think? Regardless of your initial reaction, or your considered opinion, the key question is *why* do you hold that position? What are the principles, the ethical steps, or moral paradigm that you employ to reach your decision?"

In these days of rapid medical advance we need clear thinking to answer these very practical questions. Preferably we can arrive at a "preemptive" ethical stance and avoid the retrospective question of whether or not any given procedure was morally right.

In preparation for writing these pages, I asked a number of practicing physicians, "Has an ethicist ever helped you to make a difficult clinical decision?" Every one of them answered with an emphatic "No." How very sad. We must strive to make ethical reflections useful for dealing with the dilemmas of life.

As a board-certified Ob-Gyn who specialized in infertility before it was a specialty, I encountered a wide range of reproductive issues in my practice. I was in medical school when the *Roe v. Wade* decision dramatically changed the direction of U.S. culture and its view of life and the abortion controversy. The "sexual revolution" and sexually transmitted diseases were exploding across the consciousness of America, herpes became epidemic, human papilloma virus (HPV) was linked to cervical cancer, and the elusive "chlamydia" was determined to be the culprit behind much tubal disease and resultant infertility. And then came HIV infection (with its result, AIDS) — perhaps the most frightening diagnosis since the plague — raising so many issues pertinent to its spread and treatment. And now, as HIV spreads so rapidly, pregnant women join the increasing risk population. What are the ethical ramifications of treating — or not attempting to treat — an HIV-infected pregnant woman with no good human studies to demonstrate safety for the infant? Are we facing another thalidomide or DES catastrophe?

I was in training when the first so-called test tube baby was successfully conceived. Pregnancy, contraception, and family planning, natural or otherwise, became daily features of life. Controversies, even among Christians, became increasingly vocal. Can Christians use contraception? Can one hold a high view of the sanctity of human life and use contraceptive devices or methods? Which ones? And what about technologies to assist reproduction?

Hardly a day in practice went by without some of these issues surfacing, and the camps became deeply polarized, firmly entrenched, and armed to the teeth with the "authoritative word." What light can biblical principles and other ethical considerations shed on this debate?

As I think of the abortion issue, I must tell you of Nanette, a dear Christian patient I cared for fifteen years ago, yet remember like it was yesterday. She presented to my office as a newlywed, seeking advice regarding pregnancy and her health condition. She was in her late twenties and had had a heart valve replaced some fifteen years prior to that time. Of course, valve surgery has improved considerably since then, but the valve that she had was functioning poorly with no real hope for improvement short of replacement. I agreed with her cardiologist that she was not a good candidate for pregnancy, that the stress of increased blood volume and all the changes associated with a normal pregnancy might prove more than her heart could handle. In short, pregnancy could threaten her life.

The next time I saw Nanette, several months down the road, she was a few weeks pregnant and in trouble already. Old artificial mechanical heart valves are not well suited to pregnancy, and the changes in the blood's clotting tendency make these pregnancies very challenging under the best of circumstances. Back in those days, the effect of using anticoagulants in early pregnancy was a mystery. We did not really know how great the risk of causing a fetal deformity was, especially limb shortening, but we knew if the mother clotted and stroked we could certainly lose both mother and child. Many had advised Nanette toward an early abortion, including some medical "experts." But knowing her faith commitment, I knew that would never be an option for her.

Her cardiologist and I cautiously nursed Nanette through the pregnancy, crisis by crisis, hospitalization by hospitalization, in and out of cardiac failure, fluid imbalance, and clotting emergencies. Finally, at thirty-five weeks, about five weeks before her due date, and fairly good size even for those days in "neonatology," Nanette presented again to the hospital in early heart failure. The medical staff swarmed over her, and we got her stabilized and shortly thereafter prepared for an operative delivery. The cesarean section went off without a hitch, and her beautiful daughter, with four perfectly formed limbs, was born.

Those of you familiar with the cardiovascular changes in pregnancy understand that the body has months to adapt to the slowly increasing blood volume, weight, and stress, but with delivery, the changes are fairly dramatic in the opposite direction. Nanette was being carefully monitored in intensive care when she crashed; the valve function diminished dramatically, and the cardiothoracic surgery team decided only emergency valve replacement might save her life. We took her to surgery shortly thereafter, and as we moved her to the operating table she looked me in the eyes and said, "Dr. Bill, I'm scared."

When they removed the old valve, the opening was no bigger than a straw, and the nice new valve looked great, but when it was time to come off the pump, that heart would not beat. After extensive and heroic efforts to no avail, Nanette was pronounced dead. This temporal life is fragile. There were several other cases during my obstetrical years, critical life-and-death circumstances, that should give us pause before stating dogmatically, as some do, that the "life of the mother is never at risk in pregnancy." Life, though, is indeed precious, regardless of its length.

What about abnormal pregnancies, those with genetic problems or anatomic developmental problems that will affect the so-called quality of life? Here, at times, medical personnel reveal their bias. I consulted on a case of a woman who was discovered to be carrying a child with Down Syndrome. The child also had a rather dramatic condition where much of the abdominal contents develop and remain outside the body. The mother was told her child would be a "vegetable" and never be capable of any "human response" — and thus she should terminate the pregnancy. Now, I'm not suggesting that raising a child with Down Syndrome is ever simple and easy. However, it is impossible to predict how this baby will do, and the anatomy problem is surgically correctable. Decisions of this magnitude must be based on accurate information, at least to the extent that such information can be known. *Doctors can give bad advice based on personal opinion and bias.*

I still receive calls about ectopic (tubal) pregnancies, that is, pregnancies in which the zygote implants outside the womb. Doctors have advised surgery to remove the pregnancy, and the mother is often torn about the implications of this procedure as an abortion. Some well-meaning pastoral advisors have added to the problem by suggesting that the baby can be "moved" or that the baby might "migrate" and be all right. At this moment, we have no way to achieve such movement; the ectopic pregnancy still represents a great risk to the mother's life. Women die in the United States today because of undiagnosed ectopic pregnancies that rupture. The mother bleeds to death before surgical intervention can save her. We must be accurate in our understanding if we are going to advise people in the medical arena. Our zeal for the life of the baby must not dampen an informed zeal for the life of the mother. *Pastors and well-meaning Christian friends can give bad advice based on wrong information or faulty understanding.*

Consider also the rapidly expanding horizon of infertility diagnosis and treatment: assisted reproductive technologies. We live in very exciting times. Over the past several decades the advances have been enormous. Not long ago, Chlomid was virtually the entire armamentarium for assisting couples to conceive. Now with micromanipulative techniques a single sperm can be inserted into a single egg with a reasonable chance for fertilization and development of an embryo. Remarkable! But a word of caution is in order, for just because we *can* do something does not necessarily mean we *should* do it. Each medical procedure must first pass through the grid of theologically sound, ethical thinking. With each advance comes a commensurate responsibility to use the technology wisely.

Some argue that no such reproductive procedures should be employed, that God closes the womb and he can open it whenever he wants, without anyone's help. Others argue that we treat cancer with chemotherapy, infections with antibiotics, and diabetes with insulin, so why not treat infertility with appropriate measures? In fact, sometimes the treatment is simple antibiotics, hormone adjustment, or even surgery to remove a blockage. Proper diagnosis is critical in the decision-making process for both doctor and patient.

However, what about the really controversial areas, like donor insemination, postmortem sperm collection, surrogacy, cryopreservation (freezing embryos), selective reduction (abortion) for multiple pregnancies, and cloning? Are there any limits? Does anything go?

Now we reach the crux of the matter. How do we decide? What ethical principles or paradigms apply to the ever-expanding scientific horizon? Are there right answers?

We live in a day and culture in which many do *not* have as a presupposition that there is a "right answer." Many have accepted the premise that the only thing one can truly "know" is what can be scientifically tested and proven. Anything else cannot be known — there is no truth, only opinion, perspective, and paradigm. And thus Christian viewpoints have become largely irrelevant to the marketplace of opinion and debate, privatized as a separate "non-intellectual" sphere. This leads to a moral/cultural relativism, with all values determined by individual communities. We must be ready intellectually to give a defense. We must be competent and conversant in our area of expertise, whether that is medicine, philosophy, or ethics. Ultimately, many of these issues distill down to our understanding of personhood, true humanity, marriage and family.

I am reminded of a thirty-year-old patient who had experienced premature menopause. The condition is not very common, but certainly one familiar to most primary-care physicians. This particular woman had not achieved pregnancy before she ceased ovulating. She had eggs donated from her sister and inseminated in vitro with her own husband's sperm, and then reimplanted into her own uterus with total hormonal support. In this case, cryopreservation was used back when the successful thaw rates were quite unsatisfactory. A heroic loving gift on the part of her sister? Or a violation of the scriptural "one flesh" principle and a child outside the marriage?

In answering such questions, we can miss the mark in more than one way. Sadly, many people are intellectually lazy and even gullible. They are content to have someone else do the thinking, and then they will echo the resulting opinion: "Give me an expert that can tell me how to think (in twenty-five words or less)." Theologians with no medical training will diagnose and suggest treatment plans for people, or tell them with little humility what God thinks about issues on which God has not specifically spoken. Physicians or scientists with no ethical training will declare certain practices acceptable with little regard — or perhaps awareness — of the wide-ranging ethical implications involved. We can be more sure than right. Confidence can conceal ignorance.

Can we answer the question as to when individual, unique human life begins? When personhood is present? The origin of the soul? People are using terms like "pre-embryo" or "products of conception" — implying living tissue without humanity or personhood. Such discussions must be rooted in an accurate understanding of the biological process and guided by sound theological and philosophical perspectives. With such multidisciplinary tools we can effectively address questions such as freezing sperm and eggs as well as embryos, selective reduction, pregnancy termination for "genetic abnormality" or incompatibility with independent life, genetic engineering, and cloning.

*   *   *

We all need to listen, to learn, to think, to analyze, and to pray for wisdom and discernment as we struggle with these issues. There are godly, committed men and women who differ on many controversial issues. We must strive to understand, to hone our thinking, to do good science. We must set aside our preconceptions and presuppositions, and search the Scriptures. We must examine our hearts, know if we have wholeheartedly sought truth, and be clear whether our views are derived from scriptural principles, tradition, doctrinal consensus, or opinion. While pressing on with academic zeal, Christians must be diligent in their love toward one another. Without such love, the world will not recognize Christians as Christ's, and without Christ, the world will never know "the way, the truth, and the life" (John 14:6).

PART I

FOUNDATIONAL ISSUES

CHAPTER 1

Separating Sex and Reproduction: The Ambiguous Triumphs of Technology

Nigel M. de S. Cameron, Ph.D.

The coming of a new millennium has presented us with an important opportunity to reflect on the fundamental significance of technology for human values. There are other ways of construing the central question in contemporary debate within our culture, although any analysis that does not contend with underlying assumptions about human nature is seriously deficient. Indeed, it could be argued that part of our present cultural malaise lies precisely in our capacity to address questions of cultural shift merely *ad extra,* as if the settled character of human being — what it means to be a member of *Homo sapiens* — were beyond dispute. Yet the depth of our problem is indexed precisely in the degree to which changes in our cultural environment (public policy, the professions, science and technology) reflect and in turn affect our conception of who we are.

A cultural revolution could indeed be defined in such terms: as that period during which the connection between what people believe themselves to be, and the cultural institutions and precepts around them, is particularly fluid, as traffic passes both ways. On such a scale of measurement, this breakpoint generation's experience must rank at the top. If we share the illusion that the changes around us are merely around us, we participate most thoroughly in the contemporary myth that it does not matter who we think ourselves to be. If we break with that illusion, and see at the center of all the cultural transitions around us the question of human being, we gain the measure of the challenge which we confront. And this is nowhere more true than in the arena of science and technology, in its application to medicine and the new powers we have to determine ourselves.

In other words, the threat to human dignity posed by the technological reduction of human nature and its possibilities lies close to the heart of the malaise in contemporary culture. The exponential growth of technological possibilities threatens to mesmerize us and distract us from the broadest context in which these wonders have arisen, which is that of the fruitfulness — sometimes wayward and sometimes wise — of the human mind. That fruitfulness needs to be engaged wholly to the task of our potential for human flourishing. If its products have another significance, or are fraught with the ambiguity that so often attaches to our fallen creativity, then our critique must be adequate to the task.

Yet the fundamental question of the significance of technology for human values is largely unexplored. Such is the case not simply in public policy — where we tend to meet individual controversies, isolated from one another and from the general questions which are at stake — but even in the narrower context of Christian reflection. That this should be true even here, in the matter of bioethics, is a particular surprise and a matter for remedy. This can be well illustrated by the latest locus of controversy, the discussion over human cloning, for which the Christian community was as unprepared as the general population. These questions throw us back time and again on the basic issue of the meaning of our humanity.[1] Yet technology is constantly taking us by surprise. We find ourselves floundering because we face essentially unexamined questions.

The index of our predicament is found in the enormity of the misjudgments made by otherwise sound and competent people. For example, I recently received a call from an elder in a large church who sought advice. The church's deacon board was being asked to fund the medical costs of a member of the congregation who was pregnant and had been hospitalized because of preeclampsia. Yet the case had some unusual features. This was an in vitro pregnancy. The woman was an upstanding member of the choir. One of the pastors, she said, had advised her to undertake in vitro fertilization. And the reason? She had told him that (a) she wanted to have a baby, but (b) she was not married.

---

1. This is the approach that has characterized much Roman Catholic writing on sexual ethics, which it has been more clearly methodological than Protestant discussions. For *Humanae Vitae* and other representative contributions, see Stephen E. Lammers and Allen Verhey, ed., *On Moral Medicine: Theological Perspectives on Medical Ethics,* 2nd ed. (Grand Rapids: Eerdmans, 1998).

At the heart of our problem in confronting so many of these questions is that we engage them *seriatim;* we see trees and more trees, but it is dauntingly difficult to grasp the shape of the woods. Part of the value of the present volume is to lay on the table a succession of these questions in the hope that we shall be aided in our discussion of each one by our discussion of each of the others. That is to say, by seeing the conspectus of questions confronting us, we will be driven to fundamental methodological reflection on what they have in common and on the assumptions that lie behind our approaches to them.

This chapter groups together two important questions, the significance of contraception and of assisted human reproductive technologies, in order to focus the related challenges they raise for our assumptions about human sexuality. To put it differently: we focus here on the significance of the divorce of sexuality and reproduction. When the history of the twentieth century comes to be written, this divorce will surely rate among the most significant achievements of our age — a result of the problematic marriage of technology and human values according to the deleterious assumptions of the closing years of the second millennium.

The Demoralization of Society

These developments are strange, for we are the legatees of certain fixed assumptions which have influenced and determined the identity of our culture to a remarkable degree. In one of the most interesting books in recent years, *The Demoralization of Society,* Gertrude Himmelfarb reflects on the collapse of the moral structure of the West.[2] Indeed, the Achilles' heel of much contemporary conservative cultural critique lies precisely here, in confused or misstated understandings of the past. We need to be careful, since it is easy for us to generalize in a misleading fashion about the past and its supposed superiority to the present. By focusing attention on the central significance for our culture of its changing assumptions about the significance of human being we are seeking to clarify the momentous importance of what is taking place in this generation. Yet we must do so without spreading myths about the past. Some things are much, much better now than they were; others are worse and degenerating into chaos. Candor in this exposition and analysis will only help our case. The fact that the civil rights movement and many aspects of feminism have brought new dignity and a proper freedom to our generation does

---

2. Gertrude Himmelfarb, *The Demoralization of Society: From Victorian Virtues to Modern Values* (New York: Knopf, 1995).

not need to be discounted, especially if it is recognized how much these changes owe to the seeds sown long ago in the Judeo-Christian vision of human dignity.

In her book, Himmelfarb points out some of the other dynamics involved here and documents her claims in an unusually detailed manner. Let us look at one example. She draws together revealing statistics which illuminate the huge change in sexual practice that has marked the twentieth century and has set it in dramatic contrast to the nineteenth. She writes:

> In nineteenth-century England, the illegitimacy ratio — the proportion of out-of-wedlock to total births — rose from a little over 5 percent at the beginning of the century to a peak of 7 percent in 1845. It then fell steadily until it was less than 4 percent at the turn of the century. . . . In East London, the poorest section of the city, the figures are more dramatic because more unexpected; illegitimacy there was consistently well below the average: 4.5 percent in midcentury and slightly under 3 percent by the end of the century. Apart from a temporary increase during both world wars, the ratio continued to hover around 5 percent until 1960. It then started a rapid rise: to over 8 percent in 1970, 12 percent in 1980, and then, precipitously, to more than 32 percent by the end of 1992 — a two-and-a-half times increase in the last decade alone and a sixfold increase in three decades. In 1981 a married woman was half as likely to have a child as she was in 1901, while an unmarried woman was three times as likely.
>
> In the United States, the figures are no less dramatic. Starting at 3 percent in 1920 (the first year for which there are national statistics), the illegitimacy ratio rose gradually to slightly over 5 percent by 1960, after which it grew rapidly: to almost 11 percent in 1970, over 19 percent in 1980, and 30 percent by 1991 — a tenfold increase from 1920, and a sixfold increase from 1960. . . . For whites alone the ratio went up only slightly between 1920 and 1960 (from 1.5 percent to a little over 2 percent) and then advanced at an even steeper rate than that of blacks: to almost 6 percent in 1970, 11 percent in 1980, and just under 22 percent in 1991 — fourteen times the 1920 figure and eleven times that of 1960. If the black illegitimacy rate did not accelerate as much, it was because it started at a much higher rate: from 12 percent in 1920 to 22 percent in 1960, over 37 percent in 1970, 55 percent in 1980, and 68 percent by 1991.[3]

There are many lessons to be drawn from these remarkable statistics. At a time of widely unreliable contraception, illegitimacy was held down — only

3. Himmelfarb, *Demoralization of Society,* pp. 223ff.

to take off with explosive force at the very moment when "birth" came seemingly under "control" with the oral contraceptive. Illegitimacy had actually *decreased* through the nineteenth century, not least because of the spread of religious values. Counterbalancing the undoubted benefits of social and cultural change in recent decades, we have to note one solid fact: that it was very rare for children to grow up outside a settled context of family life in nineteenth-century Britain and the United States. Without even considering the parallel explosion in divorce and family breakdown, we must note this radical shift in their — our — experience. Undergirding sexual mores and their implications, the old moral structures of our culture have collapsed — producing many of our current problems. Without a naive golden age view of the past we can very properly look back to a time when assumptions were different and therefore conduct was different too.

This analysis raises a broad question of method to which we can only allude at this point: that it is difficult to understand anything in the present cultural context in Western civilization without seeing it either as the direct lineal descendant of the ancient values of this amalgam of classical and Judeo-Christian civilization which we call "the West," or as a denial of these values. Accordingly, Christians and Jews often share a more informed way of reading what is taking place in the culture than those who have no sympathy for its past, since it is in the interest of secular and post-Christian interpreters to deny equally the continuities and discontinuities as they seek to interpret the present. Deep in the fundamental self-understanding of human nature in our culture, we discover the essential conjunction of sexuality and reproduction. The opening chapters in the Bible — Genesis 1, 2, and 3 — reveal the manifesto for so many of the cultural components of sex, marriage, and family that have flourished within the history of our civilization. In Genesis 1 there is the command to multiply, followed in Genesis 2 by the first institution of marriage as Jesus Christ himself refers to it — the "leaving and the cleaving," and the resulting union. We see Genesis 1 and 2 together as if we are overlaying slides one on the other. We see unity and procreation together defining our assumptions about the meaning of sex and of reproduction, and their focus in marriage.

The Great Bifurcation

However, our theme here is not marriage itself, but the more raw question of sexuality and reproduction. As we look back on the history of the twentieth century, we see the way in which sexuality and reproduction have begun to be bifurcated, with dramatic implications for both. For two generations now it is generally agreed that effective contraceptive has been available. While we might wish to raise questions about effectiveness, properly used techniques can generally prevent conception as an outcome of sexual intercourse. Whether oral contraceptives, condoms, "natural family planning," or other methods, contraceptive techniques have increasingly given people the capacity to make decisions about how in their own relationships they will relate sexuality and procreation.

On the other hand, and with more drama, we now have the range of questions raised by assisted human reproduction. Each form of "assistance" is essentially an example of the way in which veterinary medicine is taking over the assumptions and techniques of human medicine. A parallel development is taking place in end-of-life discussions.

There have been three stages in the growth of assisted reproduction. First, two generations ago there were debates as to whether artificial insemination was acceptable, particularly AID, the use of donor insemination, as opposed to AIH. (The latter term has now moved from being "AI by husband" to the more politically correct "AI homologous" — in fact, the more recent shift to the single term "intrauterine insemination" has sometimes glossed over the husband-donor distinction entirely.)

The controversies of two generations ago largely form the context in which the second-stage discussions about in vitro fertilization took place, and they explain part of the reason the discussion of in vitro fertilization was so cursory and did not have a serious impact on Christian perception of these techniques. In vitro fertilization has largely been absorbed by the church, not least by the evangelical church, and it is hard to engender serious discussion since so many Christians have failed to engage in a theological critique of contemporary challenges to the notion of human value and the significance of technology.

Human cloning marks the third stage, and again many people are saying, "Well, cloning is the same kind of thing as in vitro, and we have already had the conversation." It is interesting the extent to which cloning makes both intrauterine insemination and in vitro fertilization look "natural," because cloning threatens a great third wave which will take us much more dramatically out to sea in a post-sexual view of human reproduction.[4]

4. The question of cloning and its potential application to humankind is well discussed in Leon Kass and James Q. Wilson's essays brought together in *The Ethics of Human Cloning* (Washington, D.C.: The AEI Press, 1998). Kass argues forcefully against such a practice in his paper "The Wisdom of Repugnance"; Wilson takes another and disturbing tack, in which individual liberty is given center stage.

Taken together, these various forms of assisted reproduction all achieve the essentially unintended result of giving us a way to reproduce without sex. The single woman described earlier in this chapter, who became pregnant through in vitro fertilization, is a good example of somebody who has completely divorced sexuality from reproduction and managed to believe she could do so within the context of faith. Here is a woman who does not want to get married, yet who could not bear the thought of finding herself in an improper sexual relationship. She seeks advice and is told, "They have this great technique. Go do it." And so collapses the moral integrity of the Christian tradition in the face of unexamined technology.

With the semi-perfection of contraceptive technique, then, we have disjoined sexuality and reproduction, such that we can now have what has been called "recreational sex." At the same time, we have the possibility of technological reproduction. The bifurcation is complete, and while it remains the case that most human conceptions take place in the context of sexuality, we have in principle now technically disjoined the two such that these tracks are diverging; their interconnection will become increasingly a matter of choice.

When we set these developments side by side with the implications of the human genome project and its sophisticated capacities to enable us to control and shape our own kind, we see a fundamental break that the human race has made with its biological identity. Observing this is not itself to comment on whether genetic and reproductive techniques have proper as well as improper uses. Such discerning critique is the fundamental challenge of our engagement with technology. There may be no necessary impropriety in these techniques if they are placed within the context of a proper vision of the identity of human nature and sexuality and its reproductive potential. But once we abandon that matrix within which to view the possibilities of these technological aids, then not only will they undermine our existing vision of the identity of human being and the proper place of sexual and reproductive capacities, but they will catalyze the broader process of cultural change in which those assumptions have been central. The manner in which these discussions within the Christian community are finally resolved will be of primary significance for the survival of that community among the post-Christian structures of the consistently secularized culture and its institutions.

Our failure to engage a judicious critique of technology and its implications for human nature is of the greatest significance. Unless we can address that discussion, not only will we fail to prudently disentangle the interconnected questions involved, but we will also be unable to address the more demanding questions which will arise as the human genome project, with its capacity for the wholesale reinvention of human nature, comes to center stage.

We will be relegated to hearing of these things simply through phone calls from pastors about *faits accomplis* within the Christian community.

This brief survey of the combined significance of the twin developments in the capacity of the human race to determine its reproductive powers is ripe with implications. Public policy in the West has its foundations in the Judeo-Christian tradition, and despite the self-conscious secularization which is now evident in most Western jurisdictions, the influence of distinctive understandings of what it means to be human continues to some degree. In general, public policy has handled contraception in three ways. Population policies in several countries have used taxation strategies and other means to encourage childbirth, out of a concern to maintain population levels at a time of falling birthrates. Meanwhile, population policies in many less developed nations have sought to limit the growth in population. One controversial feature of international policy in this regard has been the support of population programs (often involving abortion provision as well as contraception) by international agencies and, one step removed, Western governments. Finally, contraceptive usage since the development of the oral contraceptive more than a generation ago has generally been encouraged (e.g., through social medical provision), except within overtly Roman Catholic countries like the Irish Republic. In general, discussions of contraception have been pragmatic and narrowly focused; the broad cultural-historical significance of this crossing of the biological Rubicon has been conspicuously absent from all three of these contexts, except (largely by implication) in the case of conservative jurisdictions that have with diminishing success resisted public contraceptive provision.

At the same time, increasing availability of intrauterine insemination and in vitro fertilization has received comparatively less attention in public policy. Most jurisdictions essentially permit market forces to operate, with formal restrictions which have the effect of offering public policy approval to the key technical options and thereby to their radical assumptions about sex, marriage, and the nature of human life.[5]

A recent case well illustrates the contemporary mindset, as the divorce of reproduction[6] and sex now offers a new frame of understanding. Adding to

5. The major international approach to these questions is found in the Council of Europe's Treaty on Biomedicine and Human Rights, which is to be commended for seriousness and the clarity of its moral framework within which individual states may adopt contrary positions.

6. The use of this term, in many ways an innocent synonym for procreation, has come to have ideological overtones, since the new context is one in which the impersonal connotation of "reproduction," the language of the print shop, has come to replace "procreation" with all its sexual and theological ethos.

a welter of litigation in several jurisdictions over frozen embryos (as the genetic parents die or divorce), the Wallis-Smith case focuses sharply on the sexual act itself, and asks the law to sanction the view that the sexual act can be entirely dissociated, as an act of interpersonal pleasure, from procreative possibility. Peter Wallis and Kellie Smith "met at work, fell in love, shared an apartment. But when Smith discovered she was pregnant, the fairy tale ending didn't come." Instead, Wallis is suing Smith for having become pregnant against his will and has accused her of "intentionally acquiring and misusing" his semen when they had sexual intercourse. She claims the pregnancy was accidental, and that she was on the Pill; he claims that it was deliberate on her part, in that she stopped taking the Pill so that pregnancy would result. In its report on the case, the *Washington Post* notes that "the new frontier" of the in vitro process "has altered society's view of sperm, introducing it as something that can be bought and sold in an ever-changing marketplace."[7]

7. *Washington Post*, 23 November 1998, p. 1.

CHAPTER 2

A Child of One's Own:
At What Price?

Gilbert Meilaender, Ph.D.

At the heart of many of the moral puzzles we face, and certainly at the heart of some of the puzzles about the use of new reproductive technologies, is the "nature" of human nature. We experience ourselves as beings who are located — and yet, who transcend any given location. In biblical terms, we are created from the dust of the ground as finite beings who are located in time and space — but we are simultaneously bodies into whom the Spirit of God has breathed, bodies made ultimately for communion with God.

The story of modernity is, in large measure, a story of breaking free from the limits of what is "given" in our finitude. It is the story of freedom, which often seems to be a freedom without limit, the freedom of self-creators. New reproductive technologies, manifesting as strikingly as they do the exercise of that freedom, force us to ask about limits. Are there projects we are able to undertake, projects that might even in some sense accomplish good, which, nevertheless, we ought not endorse? Are there circumstances in which we should not use our freedom to bypass, override, or reshape the "givens" of our nature? There are, no doubt, a number of ways to come at this question in the context of new reproductive technologies. This chapter will do it by reflecting upon what it might mean to want, as we often say these days, "a child of one's own."

It is probably natural to want a child of one's own. Is it also good? Perhaps if it is truly natural, in accord with our created nature, it must also be good. But the seemingly innocent desire to have "a child of one's own," combined with the high-tech possibilities of modern medicine and the ever-present pursuit of commercial gain, has fashioned a world in which we regularly create moral conundrums that are beyond our ability not only to solve but even to name. The things we are willing to do tell a story — a story about the point of having children.

The Story Our Society Tells

Consider the following cases, all roughly adapted from true stories, chosen almost at random:[1]

- A woman unable to have a child "of her own" had her ovum fertilized with her husband's sperm in the laboratory. The resulting embryo was then implanted in the womb of the woman's mother, who, having carried the pregnancy to term, gave birth to her own grandchild.
- A husband and wife who thought they wanted a child "of their own" contracted for the conception of a child who would be conceived from sperm and ovum that came from anonymous donors and who would then be gestated in the womb of a hired surrogate. Shortly before the child was born, the husband and wife who had wanted this child divorced. A judge felt compelled to rule that the baby girl actually had no legal parents at all.
- A woman undergoing infertility treatment in order to have a child "of her own" conceived triplets. For medical reasons she was advised that it would be safest if she were to undergo "fetal reduction" — that is, reduce by abortion the number of fetuses she was carrying to one. She did, but weeks later, having undergone amniocentesis, she learned that the one remaining fetus had a genetic anomaly. She therefore aborted that fetus as well.

1. Most of these examples are drawn from actual cases. See the following sources: (1) Gina Kolata, "Reproductive Revolution Is Jolting Old Views," *New York Times,* 11 January 1994; (2) Joan Beck, "Careless Intentions," *Chicago Tribune,* 1 February 1998 (an appellate court later overturned the lower court and ruled that the couple who hired the surrogate are the baby's parents [see "Couple Ruled to Be Test Tube Baby's Parents," *Chicago Tribune,* 12 March 1998]); (3) New York State Task Force on Life and the Law, *Assisted Reproductive Technologies,* April 1998, p. 129; (4) *Assisted Reproductive Technologies,* p. 127; (5) Raymond Hernandez, "Court Blocks Embryo Use over Ex-Husband's Rights," *New York Times,* 8 May 1998; (6) Judy Peres, "Children from Beyond the Grave," *Chicago Tribune,* 15 December 1997; (7) Gina Kolata, "A 63-Year-Old Gives Birth; What Does That Say About Life?" *New York Times,* 27 April 1997.

- An infertile married couple desiring a child "of their own" underwent in vitro fertilization (IVF) and conceived a child. Four and a half months into the pregnancy they learned from amniocentesis that the child they had wanted so badly and worked so hard to make had Down Syndrome. Having learned that, they decided to abort.
- An infertile married couple sought in vitro fertilization in hopes of producing a child "of their own." Before undergoing the procedure, the couple signed an agreement saying that the resulting embryos could not be used without the consent of both parties and that, should they divorce, ownership of the embryos would be decided either in a property settlement or through a court decision. Nine attempts to implant embryos fertilized in the laboratory failed to result in a pregnancy that could be carried to term. Four embryos were implanted in a surrogate, the woman's older sister, but that procedure also failed. Shortly thereafter the couple divorced and the woman sought a court order giving her sole custody of the embryos so that she could try again to have a child "of her own." Given the prior agreement the couple had made, the court ruled that the woman could not do this without the consent of her former husband.
- A young woman about to undergo chemotherapy for leukemia but hoping nevertheless some day to have a child "of her own" had her ova harvested and fertilized with donor sperm before treatment; the resulting embryos were frozen. After she died of leukemia at age twenty-eight, her parents sought a surrogate who would agree to gestate the embryos. In this search they used the Internet and an appearance on the Oprah Winfrey show, intending that their son and daughter-in-law would raise the child if the pregnancy could be successfully carried to term.
- A sixty-three-year-old woman, wanting a child "of her own" had implanted into her hormonally primed uterus an embryo made in the laboratory from her husband's sperm and an ovum from a younger donor. She then completed the pregnancy and gave birth to a child.

Such cases could be multiplied almost without end, and we may sometimes find it hard to remember or believe that the first "test tube baby" was born just over twenty years ago. Two decades later we live in a world in which a woman can give birth to her own grandchild; in which a child can have as many as five parents (the donors of sperm and ovum, the surrogate who carries the child during pregnancy, and the two "rearing parents"); in which people can "have children" posthumously; in which parents can go to great trouble and expense to conceive a child whom they then abort if prenatal diagnosis shows that the child is "defective" in some way; in which quite soon it may be possible to give birth to identical twins born years apart; and in which it may soon be possible for a woman without ovaries to receive an ovary transplant from an aborted fetus, making that fetus the genetic mother of her child. And none of this comes cheaply. A recent report of the New York State Task Force on Life and the Law notes that "[c]onservative estimates place the cost of a successful delivery via IVF at more than \$40,000."[2]

Taken together these cases display the story we have begun to tell each other about the meaning of children. The story line is, roughly, as follows: Because having children is something many people want for their life to be full and complete, and because it is such a fundamental aspect of human life, we ought to use our skills to help them achieve that desired fulfillment. Indeed, having children is an entitlement to which there are few limits. Of course, we ought not exercise this right in a way that directly harms children, but in many cases, after all, the children would not even exist were it not for the use of new reproductive technologies. If the suffering that infertility brings can be relieved, and if children are not harmed, then high-tech reproductive medicine is a good thing. This is the story that, more and more, we tell ourselves in this society.

Consider how different an image of the child emerges in Galway Kinnell's poem "After Making Love We Hear Footsteps."[3]

> For I can snore like a bullhorn
> or play loud music
> or sit up talking with any reasonably sober Irishman
> and Fergus will only sink deeper
> into his dreamless sleep, which goes by all in one flash,
> but let there be that heavy breathing
> or a stifled come-cry anywhere in the house
> and he will wrench himself awake
> and make for it on the run — as now, we lie together,
> after making love, quiet, touching along the length of our bodies,
> familiar touch of the long-married,
> and he appears — in his baseball pajamas, it happens,
> the neck opening so small
> he has to screw them on, which one day may make him wonder
> about the mental capacity of baseball players —

---

2. *Assisted Reproductive Technologies*, p. 423.
3. Galway Kinnell, *Mortal Acts, Mortal Words* (Boston: Houghton Mifflin, 1980), p. 5.

and flops down between us and hugs us and snuggles himself to sleep,
his face gleaming with satisfaction at being this very child.

In the half darkness we look at each other
and smile
and touch arms across his little, startlingly muscled body —
this one whom habit of memory propels to the ground of his making,
sleeper only the mortal sounds can sing awake,
this blessing love gives again into our arms.

What makes this a better image than the one emerging from the examples
with which this chapter began? What story of the meaning of a child under-
lies this image? One way to think about such questions is to reflect upon the
desire to have "a child of one's own." This desire, which is simultaneously
quite natural *and* problematic, needs examination.

The Story Christians Tell

Christians too have a story to tell, a story we regularly teach to our children —
of an infertile woman who deeply desired a child of her own, how her wish
was granted, and what she then did. It is the story of Hannah, her husband
Elkanah, and their son Samuel. Why did Hannah want a child of her own? In
part, it seems, it was because she suffered the scorn of Peninnah, Elkanah's
other wife, who had children. But that only presses the question a step further. Why should this be an occasion for scorn? What is so important about
having a child? Why do people care so deeply?

When we ask such questions today, sometimes answers of the following
sort are given: "I desire the experience of pregnancy and childbirth." "I want
the experience of child rearing." "Having a child is an important part of defining who I am." No doubt there is some truth about us buried in such answers. There are deep psychological, and even biological, imperatives at work
in the impulse to give birth. But such answers, which make of the child a
means of meeting our needs, cannot be satisfactory. To think that way is already to begin to think of children as products made to satisfy some of our
desires. And, of course, if and when the product turns out not to satisfy us, we
may be hard pressed to muster the kind of unconditional love children require if they are to flourish. That, nevertheless, we do often learn to love our
children unconditionally suggests that the experience of child rearing teaches
us about something more important than our own identity.

There are, though, deeper and better reasons for having children. We would make a little moral progress were we to say, "I want a child because I want a link to future generations." Surely that was part of Hannah's desire. In her world the link between the generations may have had greater economic importance than it now does, but even in our world it is of considerable human significance. We are not angels or free spirits who can choose to be whatever we wish; rather, we are embodied creatures, located in a particular time and place. In part, at least, it is lines of kinship and descent that identify us, even though we never choose our particular location. To learn to affirm and give thanks for our place in the world is part of growing up — and, more important, part of learning how to receive the mysterious gift of life. It is, therefore, quite natural that we should want to give life even as we have received it. That takes us some considerable way beyond the narcissism of wanting a child simply as a means to fulfilling ourselves.

But it does not take us quite far enough, for this attitude toward children continues to view a child simply as *my* own, still part of the project by which I make my way in the world. We get much closer to a satisfactory understanding if we think of a child of *our* own. Elkanah is already a father, but he and Hannah together — as one flesh — are not parents. Even in so ancient a story as this one, there are hints that this too is part of the reason for wanting a child. We are specifically told that Elkanah loved Hannah. He himself tells her that she is more to him than ten sons. *Their* love-giving has not yet been life-giving, however. It is natural that they should want a child, for that child would be the sign that the love by which they give themselves to each other is creative and fruitful.

Indeed, this last step — in which they seek a child not of his own or her own but of their own — begins to take them still farther. It presses almost toward elimination of that little word "own." In the passion of sexual love a man and woman step out of themselves, so to speak, and give themselves to each other. That is why we speak of sexual "ecstasy" — a word that means precisely standing outside oneself. No matter how much they may desire a child as the fruit of their love, in the act of love itself they must set aside all such projects and desires. They are not any longer making a baby of their own. They are giving themselves in love. And the child, if a child is conceived, is not then the product of their willed creation. The child is a gift and a mystery, springing from their embrace. They could and should, if they think the matter through, quite rightly say that they had received this child as a gift of God, as the biblical writer says of Hannah: "the Lord remembered her."

Samuel is neither Elkanah's "own," nor Hannah's "own," nor even "their own." He is "God's own" — asked of the Lord and given by the Lord. He is not, therefore, simply Hannah's or Elkanah's to hold onto; rather, he must be offered back to God, as Hannah does. Lent to the Lord, for as long as he lives.

Christians, then, do not underestimate the sheer human significance of biological ties. We understand the deep desire to have children. But we must also constantly remind ourselves that children are not our possession; they are gifts of God. They exist not simply to fulfill us but as the sign that, by God's continued blessing, self-giving love is creative and fruitful. And what if the Lord does not "remember" us as he remembered Hannah? That is reason for sadness, but it is not reason to take up the "project" of making a child. The couple who cannot have children may adopt children who need a home and parents, or they may find other ways in which their union can turn outward and be fruitful. They, too, will receive gifts from God — different gifts.

Living Our Story

If this is how Christians understand the meaning of the presence of children, how shall we evaluate the vast array of new reproductive technologies — not, for the moment, as a matter of public policy, but simply as possibilities within our own lives?

The first thing to note is that many of the new techniques involve parties other than husband and wife in the reproductive process. (This is usually the case because the couple is infertile. There are also circumstances, however, in which fertile couples might turn to assisted reproduction techniques — for example, if one of the spouses carries a serious recessive genetic disorder. The moral issues remain essentially the same, however.) Artificial insemination and in vitro fertilization very often involve sperm and egg from anonymous donors, and there is an irony here that we should not ignore. If what infertile couples want is a child "of their own" in the genetic or biological sense, techniques using donated gametes will not provide it. They are, in a sense, deceiving themselves. In the name of having a child of their own, they fail to honor the importance of biological connection, of kinship and descent.

Imagine a case in which a married couple seeks donor insemination because of the husband's infertility. Someone might say, of course, that the child whom they produce is, at least, genetically related to the mother — it is *her* own, even if not also *his* own in the same sense. And for Christians that is exactly the cause for worry. The child is to be *theirs*, not *hers* or *his*. The deliberate and willed asymmetry of relation — so unlike the mutual symmetry that exists in adoption — is precisely the problem. This child is no longer the fruit of their one flesh union. Its genetic connection to the mother, or the opportunity it provides for her to experience pregnancy and childbirth, are her individual projects. Even if her husband also desires that connection and wants her to have the experience, he shares this project only in thought, not in the body. The child cannot be the fruit of an embrace in which husband and wife step outside themselves, their aims and projects, and receive a child as a gift, a sign that the Lord has remembered them.

If we imagine the opposite sort of case — in which the ovum rather than the sperm is donated, or in which the child is gestated by a surrogate — the same concerns will be in play. In addition, something disturbing happens to the relation of mother and child. Fatherhood, paternity, has always been a somewhat detached and "intellectualized" relation during pregnancy. Paternity is not obvious. It can be disputed. Fathers must think themselves into relation with the child in the womb. Not so with maternity. A pregnant woman need not think herself into relation with the child — she experiences that bond constantly. But once more than one "mother" — genetic and gestational — has become part of the process, maternity also becomes a disputable fact, and we have the court cases all around us to prove it.[4]

Moreover, Christian spouses who set foot on what was once a back road but has now become an interstate highway of assisted reproduction should know how difficult it may be to find an exit ramp once they have begun this journey. They are not really seeking to correct a medical problem but to bypass it in order to satisfy a deep and very important desire — to have a child. But if it is the couple's desire that is being treated, we need to remember that they may not simply desire a child. They probably also desire, for example, a healthy child. And new reproductive technologies more and more commonly involve genetic diagnosis of the newly formed embryo before it is implanted in the uterus. The pressure to discard embryos who do not meet desired specifications — and to try again — may be almost impossible to resist. Spouses who undertake this journey are likely to have invested many dollars and years — not to mention tears — along the way. It is only human to want the best possible result. Understandable as that is, however, it is no longer quite the same kind of unconditional love for a child who is not our product but God's gift.

There is one possible exception to the claim that Christians ought not participate in new reproductive technologies that involve sperm or ovum from third parties. Rather than using either sperm donation alone or egg donation alone, a couple might "adopt," gestate, and rear a donated embryo. In

---

4. Barbara Katz Rothman, *The Tentative Pregnancy* (New York: Penguin Books, 1987), pp. 233-38.

such a case, unlike sperm or egg donation, the child will not be genetically linked to either parent. Although this very fact may cause concern to some who think of reproductive medicine as providing new ways to get a child "of one's own," from the Christian perspective it is preferable. We might think of it as adoption that occurs before rather than after the child's birth. The relation of husband and wife to the child is symmetrical, and they do not deceive themselves into supposing that this is in any genetic sense a child of their own.

Unfortunately, at least in our society at the present time, embryo donation is not likely to be analogous to adoption. There are already a few clinics that sell embryos to infertile couples who want a child.[5] Sometimes the embryos are custom made — allowing prospective parents to choose a combination of sperm and egg donors that best satisfies them. Other times the embryos are extras that were made for an infertile couple who achieved a pregnancy without using them. Christians could, of course, understand themselves to be rescuing such children — who as spare embryos can only be implanted in a womb, used for research, frozen and stored indefinitely, or discarded. But if we are looking for needy children to rescue, they are, alas, all around us in our foster care system. Pre-birth embryo adoption is not likely to signal similar attempts at rescue. It is far more likely to be one more way of exercising quality control, of finding the child whom we want — rather than loving the child we have been given.

In short, many of the new reproductive technologies will involve the use of third parties. In so doing they break the connection between love-giving and life-giving in marriage. That is not just a minor nuance, for it is this connection that teaches us to think of the child as a gift, that keeps us from thinking of children as our project, as existing for the sake of satisfying our desires. It is no accident, then, that these technologies usually encourage genetic diagnosis — whether before implantation or after — of the "fitness" of the embryo or the fetus. If we understand the child as our project, if we accept that kind of responsibility, then we may inevitably find that "quality control" seems like an obvious — perhaps even imperative — part of the process. This is a journey we ought not even begin.

But what if no third parties are involved? There are certainly some circumstances in which an infertile couple might make use of new reproductive technologies while using only their own sperm and ova. Women may take drugs to influence ovulation. This may be combined with assisted insemina-

5. Gina Kolata, "Clinics Selling Embryos Made for 'Adoption,'" *New York Times,* 23 November 1997.

tion — when the sperm are placed directly in the vagina, cervix, or, even uterus — if the man's sperm count is low. Either or both of these may often be part of an in vitro fertilization (IVF) procedure, in which both sperm and ovum are externalized and fertilization takes place in the laboratory. It is even possible now, within the IVF procedure, to inject a single sperm into the ovum.

Even when no third parties are involved there are serious moral concerns in the use of new reproductive technologies. The couple will be encouraged to "screen" the embryos formed in the laboratory, to consider whether a particular embryo is really the child they desire. If more embryos are produced than are implanted in the woman's uterus, they will have to ask themselves what should be done with the extras. Even apart from any IVF procedure, the use of ovulation enhancing drugs alone means that the possibility of multiple pregnancies — triplets and even higher-order multiple births — is greatly increased, and such pregnancies involve significant risks for the children conceived. They are much more likely to be born prematurely and to have low birth weight, and they may suffer lifelong complications as a result. Multiple pregnancies also mean that the couple will have to deal with recommendations of "fetal reduction." In general, and even entirely apart from the use of donated sperm or eggs, it becomes increasingly difficult to think of the child as a gift and not a product. These are simply some of the hazards of the road they are traveling.

When we remember again the number of needy children who go unadopted precisely because of their needs, when we consider the degree to which new reproductive technologies have — in a very short time — begun to teach our society to think of reproduction as a right to which everyone is entitled, when we ponder the implications of these technologies for our society's understanding of children, we must ask whether Christians should not call a halt — at least for themselves. We do not have a story that teaches us to think of children as our entitlement or our possession. Indeed, the story we tell goes even beyond that of Hannah, Elkanah, and Samuel. For knowing as we do that God has already provided the Child, we can free ourselves of the feverish need to have a child of our own, whatever the cost. Perhaps the greatest service we can perform for our own children and for the world into which they will be born is to live in such a way that we remind ourselves and others that each child is indeed not our product, our project, or our possession, but a "blessing" that "love gives again into our arms."[6]

6. An earlier version of this chapter was published as "Biotech Babies" in *Christianity Today,* 7 December 1998, p. 55.

CHAPTER 3

The Goals of Medicine:
The Case of Viagra

Dónal P. O'Mathúna, Ph.D.

The introduction of Viagra (Pfizer's trade name for sildenafil) has been an amazing event. No drug has been dispensed at a faster rate immediately after its release. Prescribed 37,000 times during its first week (in April 1998), it exploded in popularity with 250,000 prescriptions written during the last week of April.[1] More men in the United States were prescribed Viagra during April and May 1998 than sought treatment for impotence during all of 1997.[2]

Impotence, or erectile dysfunction as it is now called, affects about 30 million U.S. men. The frequency is age-related: 7 percent of men in their twenties have partial or total impotence, while 57 percent of those in their seventies have the condition.[3] Clinically, erectile dysfunction is defined as "the persistent inability to achieve or maintain an erection sufficient for satisfactory sexual performance."[4]

Viagra is in many ways the perfect magic bullet. Not only does it work primarily *where* someone wants it to work, but it only works *when* someone

---

1. Raymond C. Rosen, "Sildenafil: Medical Advance or Media Event?" *Lancet* 351 (May 1998): 1599-600.

2. Kenneth A. Goldberg, *Viagra: The Potency Pill* (Lincolnwood, Ill.: Publications International, 1998), p. 6.

3. Raul C. Shavi and Jamil Rehman, "Sexuality and Aging," *Urologic Clinics of North America* 22.4 (November 1995): 711-26.

4. I. Goldstein, T. F. Lue, H. Padma-Nathan, R. C. Rosen, W. D. Steers, and P. A. Wickers, "Oral Sildenafil in the Treatment of Erectile Dysfunction," *New England Journal of Medicine* 338 (May 1998): 1397.

wants it to. When a man is stimulated sexually, nitric oxide in released within the penis which produces the chemical cyclic GMP (cGMP). This causes smooth muscle relaxation, which allows extra blood to swell the tissues of the penis leading to an erection. This effect is reversed as cGMP is enzymatically broken down. Viagra works by inhibiting this enzyme.

Unlike other impotence treatments, Viagra leads to an erection only when a man is sexually stimulated. Its effect occurs about 30 to 90 minutes after taking it. In 21 clinic trials involving 4,000 men with various types of impotence, 70 percent found it effective.[5] However, Viagra had no effect on sexual desire, premature ejaculation, or fertility.[6] Overall sexual performance was restored to a little below normal levels.[7] Studies do not support claims that Viagra is a sexual performance enhancer or an aphrodisiac.

Viagra also had relatively mild side effects for the most part: headache (in 16 percent of men), bluish tint to vision (in 4 percent), and lower incidences of facial flushing, indigestion, and stuffy nose.[8] However, eight deaths occurred during clinical trials, almost all in men at risk of heart disease.[9] Viagra should not be taken by anyone using organic nitrates, including nitroglycerin. During Viagra's first four months on the market, the Food and Drug Administration (FDA) received 123 reports of deaths after patients were prescribed the drug.[10] Clinical details were available for only 69 of these deaths, and some may have been completely unrelated to Viagra. The FDA also acknowledged that its method of collecting post-marketing information on adverse events and deaths allows significant underreporting. However, Viagra has not been shown to have caused these reported deaths, although many who died had heart problems. These reports have led to a ban on importing Viagra into Israel, even for personal use.[11] Nonetheless, approval for Viagra is being sought in fifty other countries, and it may be beneficial in treating female infertility.[12]

Overall, Viagra is considered to be a relatively effective and safe impo-

5. Fred Charatan, "First Pill for Male Impotence Approved in U.S.," *British Medical Journal* 316 (April 1998): 1112.

6. Goldberg, *Viagra,* p. 63.

7. Goldstein, "Oral Sildenafil," pp. 1402-3.

8. Charatan, "First Pill," p. 1112.

9. Jacqui Wise, "Sildenafil Citrate Contraindicated with Organic Nitrates," *British Medical Journal* 316 (May 1998): 1625.

10. U.S. Food and Drug Administration, "Postmarketing Safety of Sildenafil Citrate (Viagra)," 24 August 1998, available at: *http://wwwfda.gov/cder/consumerinfo/viagra/*

11. Judy Siegel-Itzkovitch, "Israel Bans Import of Sildenafil Citrate after Six Deaths in the U.S.," *British Medical Journal* 316 (May 1998): 1625.

12. "Viagra's Licence and the Internet," *Lancet* 352 (September 1998): 751.

tence treatment. However, it is not without controversy. "Its introduction has also raised important questions about the role of sexuality in society, and medicine's role in addressing this fundamental human need."[13] Moreover, "many people feel that Viagra is just the first of many medications that are intended for quality of life issues."[14] These other foundational issues will be the focus of this chapter.

The Goals of Medicine

The development and use of Viagra represents a new way in which medicine can change people's lives. To a great extent, interest in Viagra stems from the reproduction revolution and changes in people's understanding of sexuality and the goals of medicine. Some see this development as the medicalization of male sexuality.[15] The use of Viagra leads to reflection on whether medicine should be concerned about relieving impotence or improving sexual experience. If the development and provision of Viagra falls within the scope of medicine, does it matter how it is used? Are physicians justified in prescribing Viagra for some cases of impotence, and not for others? Should insurance companies pay for Viagra if patients claim it improves their overall health? Are there particularly Christian issues that should influence decisions to use or prescribe Viagra?

Leon Kass has described the broader need to determine the goals of medicine: "All kinds of problems now roll to the doctor's door, from sagging anatomies to suicides, from unwanted childlessness to unwanted pregnancies, from marital difficulties to learning difficulties, from genetic counseling to drug addiction, from laziness to crime."[16] Viagra is just another technological development forcing people to ask what the goals of medicine ought to be. The most common and instinctive answer to this question is that medicine should seek to remove disease and promote health. However, this requires definitions for two other complex concepts: disease and health.

The difficulty of defining these concepts can be exemplified as follows. Probably the best known definition of health is in the 1948 Constitution of the World Health Organization (WHO): "Health is a state of complete physi-

13. Rosen, "Sildenafil," p. 1600.

14. Goldberg, *Viagra*, p. 65.

15. Raymond C. Rosen, "Erectile Dysfunction: The Medicalization of Male Sexuality," *Clinical Psychological Review* 16.6 (1996): 497-519.

16. Leon R. Kass, *Toward a More Natural Science* (New York: Free Press, 1985), p. 157.

cal, mental, and social well-being and not merely the absence of disease or infirmity. The enjoyment of the highest attainable standard of health is one of the fundamental rights of every human being without distinction of race, religion, political belief, economic or social condition."[17] In 1984, the WHO added spirituality to its list of factors involved in health.[18]

In contrast, others hold to health as a purely functional issue, believing that "health consists in the functioning of any organism in conformity with its natural design as determined by natural selection."[19] Between these extremes is Kass's definition: "Health is a natural standard or norm — not a moral norm, not a 'value' as opposed to a 'fact,' not an obligation — a state of being that reveals itself in activity as a standard of bodily excellence or fitness, relative to each species and to some extent to individuals, recognizable if not definable, and to some extent attainable. . . . If you prefer a more simple formulation, I would say that health is 'the well-working of the organism as a whole.'"[20]

Others claim that health and disease are whatever individuals define them to be:

> Outside the significances that man voluntarily attaches to certain conditions, *there are no illnesses or diseases in nature.* . . . The fracture of a septuagenarian's femur has, within the world of nature, no more significance than the snapping of an autumn leaf from its twig. . . . Out of his anthropocentric self-interest, man has chosen to consider as "illnesses" or "diseases" those natural circumstances which precipitate the death (or the failure to function according to certain values) of a limited number of biological species: man himself, his pets and other cherished livestock, and the plant-varieties he cultivates for gain or pleasure.[21]

This latter view is becoming more prevalent within holistic health care. Thus, the essence of nursing is said to be helping patients "determine the unique meaning that health, illness, suffering, and dying have for them as in-

17. World Health Organization, *Constitution,* available at *http://who-hq-policywho.ch/*

18. Duncan Vere and John Wilkinson, "What Is Health? Towards a Christian Understanding," in *Christian Healing: What Can We Believe?* ed. E. Lucas (London: Lynx, 1997), pp. 59-84.

19. C. Boorse, quoted in Arthur L. Caplan, "The Concepts of Health and Disease," in *Medical Ethics,* ed. R. M. Veatch (Boston: Jones and Bartlett, 1989), p. 56.

20. Kass, *Natural Science,* pp. 173-74.

21. Peter Sedgwick, "Illness — Mental and Otherwise," *Hastings Center Studies* 13 (1973): 30-31.

dividuals."[22] Since people differ in their views of health, individuals themselves should determine what treatments and therapies best promote their health. "[P]eople engaging nurses should be the teachers about what is helpful and what is not when it comes to health and quality of life."[23] Within this context, if people believe Viagra will promote their health, defined broadly, they should have access to the drug.

Definitions of health, disease, and medicine, have very broad and practical implications:

> If health and disease are nothing more than socially determined, culturally mediated and individually subjective concepts, then there will be little if any possibility of either placing medicine on a firm scientific footing or of finding consensus among experts and patients as to the proper limits of medical concerns. . . . Since so much money is spent in our own country as well as other nations on health care, and since there is so much controversy about the proper scope and responsibility of medicine in managing a host of human ailments that range from smoking, drinking, and obesity to infertility, appearance and eligibility for a broad spectrum of social benefits, the determination that health and disease are nothing more than subjective concepts whose meanings change depending upon political, economic, and cultural exigencies would have reverberations far beyond the realms of the philosophy of medicine.[24]

Many of the dilemmas arising in health care today stem from the widespread belief that good health, as defined by the individual, is the most important thing in life. According to this view of health, when men are confronted with impotence, medicine should provide whatever will restore their sexual function. The same applies to couples facing infertility. This view of health is reinforced by the high value placed on autonomous individualism in today's society.

Christians need to be very conscious of God's view of health and medicine, and to act in accordance with that view as it is presented in the Bible. Health is a broad concept. It deals with normal levels of functioning. It also involves values and quality of life. But each of these aspects must be balanced against one another, and with other goals and values.

22. Heleen Van Weel, "Euthanasia: Mercy, Morals and Medicine," *Canadian Nurse* 91 (September 1995): 35.

23. Gail J. Mitchell, "Questioning Evidence-Based Practice for Nursing," *Nursing Science Quarterly* 10 (winter 1997): 154-55.

24. Caplan, "Concepts," pp. 60-61.

Health and Values

While describing health frequently, the Bible does not abstractly discuss its meaning or the goals of medicine. Old Testament words for healing generally mean to restore something to its original condition, or make it whole again.[25] The Hebrew term *rapa'* is used for repairing a broken jar, healing a person, and restoring the nation of Israel. Other Hebrew words for healing mean to restore *shalom* to relationships, or to experience God's blessedness.

The New Testament discusses health and healing much more, but again, not just in the context of physical healing. The Greek words interweave many different dimensions of healing. The term *sozo* means to save, heal, and make whole.[26] Even a cursory examination of the healing miracles of Jesus shows that these were not just physical events, but dealt with all aspects of the person. The classic example is the Good Samaritan, who bandaged the injured man's wounds, accompanied him to the inn, and cared for him (Luke 10:30-37). He took care of the man's financial needs and returned to follow up on his progress.

In other words, the Bible affirms a holistic view of health. The WHO accurately stated that health is much more than the absence of disease. It is wholeness. But the Bible's outlook differs from both secular and New Age views of holistic health. Biblical health begins with the restoration of a proper relationship with God. Prior to this, people are alienated from God and dead in their sins (Eph. 2:1-2). People cannot be wholly healthy until they are reconciled to God, which only occurs through Jesus Christ (John 14:6). That is why he is the only door to the abundant life (John 10:10). He is the only one who can restore true life and complete health.

This holistic view of health means people should affirm the pain and loss that accompany impotence (and infertility). These should not be glibly dismissed. Health is not just a matter of avoiding diseases. When people are not complete, pain is involved. Men struggling with impotence are suffering the loss of a God-given gift. Sexuality is something wonderful to be welcomed and nurtured (within the appropriate context). A married coSuple whose ability to have intercourse is diminished or lost should be empathized with and comforted. Although only the couple may be aware of the problem, its impact can be as painful as losing the ability to see or walk. Viagra has at least raised awareness of this painful condition that affects millions of people. Hopefully more will receive help as a result.

25. Michael L. Brown, *Israel's Divine Healer* (Grand Rapids: Zondervan, 1995), pp. 28-31.

26. Leonard I. Sweet, *Health and Medicine in the Evangelical Tradition: "Not by Might nor Power"* (Valley Forge, Pa.: Trinity Press International, 1994).

Affirming the pain of impotence does not thereby justify all attempts to relieve that pain. Joni Eareckson Tada demonstrates that people of faith can move beyond their limitations.[27] People who find themselves blind or confined to wheelchairs can (and in many ways, must) learn to move on with life within these new limitations. This can certainly be difficult, and may take some time and counsel, but it is healthy to accept those limitations that come with life and aging. Perfect health cannot be attained in an imperfect world.

Today's culture promotes the belief that all dreams can be fulfilled and all desires met. This can make coping with impotence and infertility even more difficult than it has to be. As one urologist states: "The man with an erection problem is a man in serious trouble. . . . His trouble stems not primarily from a penis that is not working up to expectation, but rather from the heavy symbolic baggage that he, and all of us, attach to the male organ."[28] With the glorification of sex, a man's identity is linked to his ability to perform sexually. Impotence can be devastating for a man because of the cultural value placed on sexual performance. Values do get incorporated into views of health, making them somewhat culturally relative.

This realization also brings with it a certain degree of hope. If the severity of the problem of impotence is to some extent culturally determined, then its impact can be changed even if the physical condition cannot. If the severity of impotence's impact is to some extent based on values, changing those values can bring relief. While I myself have not struggled with impotence or infertility, God has helped me cope with other physical limitations. Ten years ago my definition of health meant being able to run 80-100 miles a week as a competitive runner. An injury brought that to an end. Today I feel healthy if I can run about 20 miles a week. My definition of health has changed, which represents significant growth in my dependence on God and his values. I did not want to change how much I ran because of goals I had which were very important to me (and tied into my identity). The pain and loss of not achieving those goals were intense. But I was comforted by God as the values underlying my life's goals were changed by his power. I came to see that my identity and purpose in life had been wrapped up in running and the achievements I expected to gain through winning races. But these were fragile, temporal goals. When they were frustrated, I was fortunate to have others help me see

27. Joni Eareckson Tada, "The Quest for Control." Audio/videotape available from The Center for Bioethics and Human Dignity (see front of book for contact information).

28. B. Zibergeid, *The New Male Sexuality* (New York: Bantam, 1992), p. 28; quoted in Rosen, "Erectile Dysfunction," p. 498.

how my goals could change to be in line with God's values and his purposes for my life.

A similar change is needed as people grapple with the more significant issues of impotence and infertility. Sexual intercourse and children are good things, created by God. But the world is fallen and brings much frustration and disappointment. Complete health is not something everyone can have, at least not until they receive their resurrection bodies (1 Cor. 15:42-44). Meanwhile, great benefit is gained from examining the values that underlie definitions of health, as is done in the next section. Replacing values that are incompatible with biblical ones, through God's power, will bring improvements in health since doing so will allow people to avail themselves of the abundant life Jesus promised (John 10:10).

Competing Values Concerning Health

Mechanistic Bodies

The body is commonly viewed today in purely mechanical ways, often in isolation from the rest of the person. An article exploring the popularity of Viagra compared human bodies to cars: "If science insists on getting more mileage out of the engine and prolonging our lives — thus allowing us to work into our 70s — what's wrong with maintaining the chrome and fenders? 'We expect medicine to deliver that,' says Susan Coleman, president of NCI Consulting, a New Jersey adviser to drug companies."[29] A consequence of this view is neglect of the very factors which may help improve people's lives. "As the average person has come to regard his or her body as an alien 'instrument,' the worth of which is in utility or pleasure, so he or she has come to gauge the 'health' of the body according to whether or not it works or feels good. When the body malfunctions, its illness is most likely regarded in isolation from the total circumstances of the patient's life, which may in fact be contributing to the dysfunction."[30]

Recent developments in impotence treatment reflect this tendency. "The development of pharmacological agents that effectively restore erectile function, welcome as they are, have led to the penis being the focus of thera-

29. John Greenwald, "Drug Quest: Magic Bullets for Boomers," *Time*, 4 May 1998, p. 54.

30. John P. Newport, *The New Age Movement and the Biblical Worldview: Conflict and Dialogue* (Grand Rapids: Eerdmans, 1998), p. 327.

peutic attention. Regrettably, it is often forgotten that the penis is connected to a body which has a brain and that frequently the body is associated with another body."[31] Many factors interact in impotence, such as other illnesses, drug side effects, alcohol and cigarette abuse, and, most importantly, interpersonal relationships. Even in organically based cases, psychological issues usually play some role. In one study, men with sexual dysfunction had as much improvement with interpersonal skills training as did another group receiving training in both sexual and interpersonal skills, leading the authors to conclude that interpersonal problems were more the root issue.[32]

Viagra provides a convenient and simple treatment for impotence. However, it may also undermine the more complete healing often needed. *Time* magazine asked about Viagra: "Could there be a product more tailored to the easy-solution-loving, sexually insecure American psyche as this one?"[33] Medicine has been criticized for its mechanistic focus and forced to become more holistic by treating sexual dysfunction. Yet simultaneously, mechanistic solutions are sought for the holistic problems medicine must now treat.

In contrast, the Bible affirms the holistic nature of health but also emphasizes holistic treatment. The Bible does not offer simple pop-a-pill solutions for health-related issues. Instead, it acknowledges the complexities of these issues and gives concrete advice on, for example, marriage — counsel which can in turn improve sexual relationships within marriage.[34]

Acceptance of Limitations

Today's culture values overcoming limitations, not learning to accept them. Everyone lives with certain limitations, but aging brings new ones. Numerous physiological changes accompany aging and many affect sexual function, both directly and indirectly. As more people seek help for impotence, "the primary care physician and urologist [are] at risk of attributing normal age-related sexual changes to physical pathology. . . . Of particular importance are the deleterious effects of ignorance, misconception, and faulty expectations

---

31. A. J. Riley and L. Athanasiadis, "Impotence and Its Non-Surgical Management," *British Journal of Clinical Psychology* 51.2 (March 1997): 100.

32. A. Stravynski, G. Gaudette, A. Lesage, N. Arbel, P. Petit, D. Clerc, J. Fabian, Y. Lamontagne, R. Langlois, O. Lipp, and P. Sidoun, "The Treatment of Sexually Dysfunctional Men without Partners: A Controlled Study of Three Behavioral Group Approaches," *British Journal of Psychiatry* 170 (1997): 338-44.

33. Bruce Handy, "The Viagra Craze," *Time*, 4 May 1998, p. 50.

34. Ephesians 5:22-33 is a good case in point.

on the sexual functioning of the aged."[35] Treating impotence in isolation from the person's overall health may be both ineffective and lead to the neglect of other important health factors.

The increased incidence of impotence accompanying aging should, for many, be more a natural limitation to be accepted, rather than an illness to be treated. When the most frequent diagnostic indicators of impotence are adjusted for age, many men diagnosed with impotence fall within the normal range for that age group.[36] Problems arise when the sexual expectations of seventy-year-olds match those of twenty-year-olds.

Dealing with limitations can be very difficult, but again the Bible offers much advice. People cannot have perfect bodies in this life (Rom. 8:18-22). But weaknesses and inabilities should not devastate us. Paul teaches how weaknesses are opportunities for the expression of God's power (1 Cor. 12:7-10). Certainly, there is pain and grief in the loss of bodily functions. God wants people to be healthy. But rather than scrambling after every pill and potion, it can often be more healthy to accept limitations. "Life's supreme success comes not in escaping pain but in glorifying God and laying hold of righteousness. Sometimes we glorify God with our bodies more in sickness than in health."[37] Accordingly, even when we do pursue treatment, we must recognize that it will not always work as well for us as for others — and that this result is no indication that God values our welfare less.

The Right to Health

Many believe people have the right to good health and the freedom to pursue whatever is believed to promote health. This belief reflects a set of values different from those of the Bible. While health is described there as good (3 John 2), and God is described as a healer (Exod. 15:26), nowhere are Christians told they are entitled to good health or promised good health in this life. In fact, the very opposite is promised. Christians will suffer in this life, even more so because they are Christians (Luke 9:23-25; 1 Pet. 4:12-14). Health is a good which may be pursued, but out of gratitude for all we already have, not as a demand or an entitlement.

Health is of value, but it must be balanced against other values. The ultimate goal of life is not to be healthy. The purpose of life is to love, serve, and

35. Shavi, "Sexuality and Aging," pp. 711, 714.
36. Shavi, "Sexuality and Aging," pp. 711-26.
37. Sweet, *Health and Medicine*, p. 48.

glorify God, incorporating love and service of others. Health can be a means toward fulfilling this purpose, but it is not the purpose itself. Today's culture claims good health is everything (as defined in bodily functions and emotional pleasure). It thereby becomes difficult for people to accept aging, limitations, or decreased sexual activity. These changes shake people's identities because of the belief that maintaining health is the most important thing in life.

People should pursue health because their lives are given to them in stewardship by God (1 Cor. 6:12-20). We are to be living sacrifices to God (Rom. 12:1). Good health helps us fulfill God's purpose in our lives. But we can find ways to serve others and glorify God, even as our bodily health deteriorates. Because health must be balanced against other values, people should sometimes place their health at risk, or decline something that might improve their health. Every parent knows that holding a sick child through the night is more important than protecting oneself from the illness or promoting one's health with a good night's sleep. Throughout history, Christians have entered regions of famine, pestilence, and war, viewing their bodily health as less important than witnessing for Christ. Many have given the ultimate sacrifice of their physical lives because there is more to life than temporal health, and there is more to health than temporal life. Decisions about health should be made with an eternal perspective.

When faced with impotence, various values must be balanced. Healthy marriages are not based on sexual intercourse alone, and sexual problems are often based on more than bodily functions. The physiological changes accompanying aging greatly affect how people become sexually aroused. Lack of communication between married couples about these changes, and how they can adapt to serve one another better, greatly contributes to diminished satisfaction with intercourse and the incidence of impotence.[38]

While pursuing therapy, the focus should remain on how a husband can serve his wife (1 Cor. 7:3-5). By taking the focus off sexual intercourse and related performance anxiety, the couple can learn ways to serve one another sexually.[39] This shift is enough to resolve certain cases of impotence. But even if performance does not improve, the couple can draw closer in love. In spite of what the proponents of hedonism would try to impart to us, there is more to a healthy life than sex, in light of the higher value of serving in the kingdom of Christ (Matt. 19:10).

---

38. Shavi, "Sexuality and Aging," pp. 711-26.
39. Rosen, "Erectile Dysfunction."

Justice

Christians should be concerned about issues of justice with any resource. Viagra is one of many products reflecting overly broad goals for medicine. Drug companies are investing billions of dollars in research and development for products to treat baldness, wrinkles, and skin blotches. "Baby Boomers who want to stay young forever — and who desires anything less these days? — are giving the pharmaceutical industry something that very few consumer-products makers have: a growing, demand-driven market."[40] The financial pay-off for these companies is clear. Viagra sales have been projected to peak at $2.5 billion per year, and its manufacturer's stock prices tripled with its introduction.[41]

These developments should be viewed in light of the entire human community. While millions are receiving medical attention for impotence, over 20 percent of the world lives in poverty.[42] Most of these people have access to little or no health care. Resources were focused on developing and marketing Viagra while diseases ravaged developing countries. "As the WHO has observed, only 5 percent of global expenditures on health research is concerned with the needs of developing countries, which suffer 93 percent of the world's premature mortality."[43]

Justice is not only a factor in large-scale decisions about health research. Each Viagra tablet costs ten to fourteen dollars. Individuals must decide if this represents the best use of their personal resources. God is especially concerned about justice and the needs of the poor (Matt. 23:23). When seeking help for impotence, the cost of treatment must be considered, especially since God has given us the body of Christ to help with psychological and relational issues. Every goal that medicine can pursue does not necessarily reflect the pursuit of justice.

The Place of Medicine

Medicine's goal is, in one sense, to promote health. Yet health is a very broad concept. Problems arise when the goals of medicine are expanded to include

---

40. Greenwald, "Drug Quest," p. 54. The actual peak may be somewhat lower because of emerging worries about possible heart-related side effects.

41. Charatan, "First Pill," p. 1112.

42. Evvy Hay Campbell, ed., *Ethical Issues in Health-Related Missions* (Bannockburn, Ill.: The Center for Bioethics and Human Dignity, 1998).

43. Hastings Center, "The Goals of Medicine: Setting New Priorities," *Hastings Center Report*, supplement (November-December 1996), p. S18.

the promotion of all aspects of health. Viewing medicine more broadly may result in neglect of important non-medical approaches. If impotence is organically based, a medical treatment like Viagra may be appropriate. Medicine would then be treating an organic problem. But using Viagra for all forms of impotence may dissuade some from seeking the psychological or relational help they need. Moreover, the mentality underlying this use will lead to further medicalization of contemporary life.

Medicine has generally focused on the physiological factors which allow optimal health. Pursuing this organic core of health has given medicine a relatively objective set of goals. These goals can generally be stated as preventing or curing physiologically based diseases, relieving pain, and caring for those who cannot be cured.[44] However, beyond this core of health exists a broader notion of health, which includes its heavily value-laden aspects.[45] Many ethical concerns arise when medicine expands its goals beyond core issues. Cosmetic surgery, use of drugs in sports, widespread use of mood-changing medications, and genetic enhancement of human potentials are examples of controversial expansions resulting from broadening medicine's goals. Medicine is a powerful enterprise, and expanding its goals into value-laden regions carries many dangers.

The value-laden aspects of health have generally been addressed by families, congregations, and health-care professionals other than physicians. This is not to say that physicians have nothing useful to contribute to these other aspects of health; nor does it mean that physicians should not be concerned about factors such as their patients' exercise, diet, occupational stress, spirituality, and relationships. But medicine should be very careful when addressing non-physiological issues. Recommendations about spiritual, emotional, and relational issues, even for the sake of health, are much more subjective than recommendations about physiological issues.

While focusing on physiological issues, physicians and patients interact on relational, emotional, and spiritual levels. Medical professionals cannot avoid, and should not neglect, these dimensions of their care. However, if the goal of medicine is primarily care of the physiological aspects of health, these should remain the primary concern of professional interactions. Discussions of broader issues should be framed differently so that patients know the topic has moved beyond the strictly medical. In an analogous fashion, pastors

44. Hastings Center, "Goals of Medicine," pp. S1-S27.

45. E. L. Bandman and B. Bandman, "Health and Disease: A Nursing Perspective," in *Concepts of Health and Disease: Interdisciplinary Perspectives,* ed. A. L. Caplan, H. T. Engelhardt, Jr., and J. J. McCartney (Reading, Mass.: Addison-Wesley, 1981), pp. 677-92.

should not neglect or deny people's medical needs, and should have enough general knowledge to recognize medical problems. But their expertise lies in spiritual areas, and they should refer people elsewhere for care of their medical needs. Pastors, physicians, and many others can thus work together to promote health, each recognizing the value of the others and the bounds of one's own professional training and gifting.

For Christians, medicine's ultimate purpose is to play a role in furthering the purposes of God in people's lives. Medicine does not "fix problems" just for the sake of fixing them. It promotes health because God promotes health. However, medicine is not the only means by which health can be promoted. Just because something promotes health does not necessarily mean medicine should be its provider. Clean water and sewage systems contribute much to health, yet medicine is not expected to tend to the water mains. Recognizing that health involves body, mind, emotions, relationships, and spirituality does not mean that medicine has a place in meeting every need in each of those areas.

The Bible portrays a healthy community in terms of a well-working body (1 Cor. 12). The body of Christ is made up of many differently gifted members. When everyone uses his or her gifts for the benefit of others, the body grows and develops. People are made to be interdependent, not independent. Yet, while everyone is individually gifted in specific ways, all are called to do certain things. All should visit the sick and comfort the suffering (Matt. 25:31-46). All should pray for others and spread the message of salvation (Matt. 28:18-20). Christian medical professionals should share their faith with patients, but they should be clear when they are speaking in the name of medicine and when they are speaking in the name of Christ.

This biblical image shows us how medicine can fit into society. Some people are gifted and trained in specific areas. Society has entrusted the medical profession with the care of the physiological aspects of health. With so much to know about the body or the mind, one person cannot be proficient in many areas.

Professional boundaries help avoid other dangers as well, like the increased medicalization of life. Some people have made health their religion. Some look at physicians as priests, if not gods. All of this serves to distract people from the healing God offers. When people's relational and spiritual needs underlie impotence, they need God's healing more than Viagra. While Viagra can have legitimate uses, it can also play a role in promoting values contrary to biblical values. We should be aware of these values and address them. We must also prayerfully examine our own decisions about using, recommending, or prescribing Viagra. To the extent that these decisions are based on worldly values, they will not lead to the abundant health God wants for all people.

CHAPTER 4

The Moral Status of Embryos

Robert W. Evans, Ph.D.

"The most powerful of all spiritual forces," wrote Emil Brunner, "is man's view of himself, the way in which he understands his nature and his destiny; indeed it is the one force which determines all the others which influence human life."[1] In the more than sixty years that have passed since these words were penned, theologians, philosophers, and ethicists have largely assumed that "man's view of himself" could be explained by a theory of morality that set forth clearly established principles for defining what it means to be human.

All of that changed, however, when in 1983 news sources reported that Mario and Elsa Rios had died in an airplane crash and left behind two frozen embryos in a Melbourne, Australia, in vitro fertilization clinic.[2] While the attention of some was drawn to the legal quandaries introduced by the untimely death of the parents (e.g., Did the frozen embryos have property rights? Would the eventual offspring stand to inherit the considerable fortune of the estate?), others were drawn to the rather thorny bioethical predicament which had resulted from the situation. What was to be done with the frozen embryos? Should they be discarded? Should they be donated to another couple? Public concern swelled as various competing moral frameworks were brought forward in an effort to address the problem.

Since 1983 few issues in bioethics have received the attention afforded

---

1. Emil Brunner, "The Christian Understanding of Man" in *The Christian Understanding of Man,* ed. T. E. Jessop (London: Allen Ltds., 1938), p. 146.
2. George P. Smith II, "Australia's Frozen Orphan Embryos: A Medical, Legal, and Ethical Dilemma," *Journal of Family Law* 24 (1985-86): 27-41.

to the moral status of human life before birth; indeed, no treatment of reproductive medicine is complete without such a discussion. While some theologians, philosophers, and ethicists have questioned whether there is any theory of morality that can be applied to the issues of our day, continued rapid advancements in reproductive technology and perinatal medicine are creating new tensions and placing ever greater demands upon us to clarify the moral status of the human embryo. Particularly at issue is the criterion for determining whether or not an embryo has a "right to life" and therefore a life that should not be jeopardized. This chapter will review several prevailing theories in an effort to address this important matter. In order to facilitate a Christian appraisal, these theories will be classified as "non-scriptural" and "scriptural."

Non-Scriptural Theories on the Moral Status of the Embryo[3]

Self-Consciousness

One of the more enduring theories concerning the moral status of the embryo is that an "entity acquires a right to life"[4] at the moment that it becomes self-conscious. Accordingly, an embryo would not have such a right. Michael Tooley is a well-known advocate of this view. He writes: "An organism possesses a serious right to life only if it possesses the concept of a self as a continuing subject of experiences and other mental states, and believes that it is itself such a continuing entity."[5] Philosopher Peter Singer puts an even sharper point on it: "life without consciousness is of no worth at all."[6] The

---

3. For greater detail regarding these theories, see Carson Strong, *Ethics in Reproductive and Perinatal Medicine: A New Framework* (New Haven and London: Yale University Press, 1997), pp. 41-62.

4. One of the difficulties encountered in writing about a moral theory that denies the status of personhood to a developing embryo is the manner in which one refers to the same. Here, the term "entity" seems (at least to this author) rather clinical, though perhaps accurate for our immediate purposes. For sake of clarity and ease of style, "entity" and "individual" will be used interchangeably and should not be construed as making reference to a human person unless otherwise stated. Furthermore, while the "right to life" is an absolute right in matters of law and in many ethical frameworks (i.e., one either possesses a right to life or does not), in some frameworks having a "right to life" is considered a level of moral status that is "acquired."

5. Michael Tooley, "Abortion and Infanticide," *Philosophy and Public Affairs* 2 (1972): 59.

6. Peter Singer, *Rethinking Life and Death: The Collapse of Our Traditional Ethics* (New York: St. Martin's Griffin, 1994), p. 190.

self-consciousness view considers the developing embryo and fetus to be non-persons because they lack consciousness of self. Accordingly, this view condones abortion as well as research and procedures that destroy embryonic life.

This theory suffers from at least three significant errors. First, and as Strong correctly notes, this view presents a most difficult situation for infants.[7] Newborns and young infants do not possess the neurocognitive capacity required to recognize themselves as continuing entities. According to Tooley's construction of this theory, the child's lack of self-consciousness prohibits him or her from being regarded as a person. Singer concurs. He and his colleague, Helga Kuhse, suggest that a period of twenty-eight days should be allowed to elapse before a newborn is afforded a right to life.[8] During this period, parents could decide whether or not the child is wanted. If not, it would be morally permissible to kill the baby.

Moreover, this view may allow denying a right to life to children under the age of approximately two years on the grounds that few of us have any memory of our own existence prior to this age. Arguably, if one has no memory of existence, one was not consciously self-aware. Once deprived of a right to life, the killing of infants and most young toddlers becomes morally permissible.

For example, if this theory were to become normative for defining an entity's right to life, then the actions of Melissa Drexler would be considered a morally legitimate act. According to her own admission, Ms. Drexler excused herself from her high school prom dance and self-delivered a healthy six-pound, six-ounce baby boy into a toilet. She then wrapped the live newborn in trash bags, placed the wrapped child in another bag, discarded the live baby in a trash receptacle, and then proceeded to take to the dance floor for an evening of celebration.[9] However, such an act not only violates the law, but offends human sensibility. Furthermore, Singer and Kuhse's suggestion that a period of twenty-eight days be allowed to pass before a child is assigned a right to life is clearly arbitrary. There are no medical, developmental, or moral grounds for selecting a time frame of twenty-eight days before affirming a right to life. As previously noted, newborns

7. Strong, *Ethics in Reproductive and Perinatal Medicine*, p. 43.

8. Singer, *Rethinking Life and Death*, p. 217.

9. Twenty-year-old Melissa Drexler admitted to knowingly and intentionally killing her newborn son before a Freehold, New Jersey, district court as a part of her plea agreement with the prosecution in this case (20 August 1998). In exchange for her guilty plea to charges of aggravated manslaughter, Ms. Drexler was promised that she would receive a jail sentence not to exceed fifteen years.

and infants are not capable of viewing themselves as continuing entities that exist over time. And there is no evidence that an infant becomes self-conscious at twenty-eight days; indeed, a lack of self-consciousness may extend for a considerable period.

Second, this theory permits the killing of those who have lost their capacity (even temporarily) to experience themselves as continuing entities. Those who are anesthetized, comatose, psychotic, suffering from substance-related disorders, or severely mentally retarded often lack the capacity to experience or recognize themselves as continuing subjects of experiences. Indeed, a person usually lacks self-consciousness when in a deep sleep. Surely it would be difficult to seriously argue that it is morally permissible to kill a person on the grounds that he or she lacked self-consciousness and, hence, a right to life while taking an afternoon nap! However, such a possibility must be sanctioned by this theory if it is to deny a right to life to embryos and fetuses who will commonly become conscious after a time.

Third, this theory is based upon the false assumption that, once an entity attains self-consciousness, this mental state remains relatively fixed over time. However, self-consciousness is more fluid than this view allows. Patients in a delirious state of mind suffer from transient periods during which self-consciousness may become seriously compromised.[10] Indeed, by definition, the patient who is delirious must have a clouding of consciousness with a concomitant reduction in the ability to focus, sustain, or shift attention. Disturbances in memory, orientation, and awareness frequently fluctuate over the course of the day, and patients who are delirious are often quite incoherent. Under the provisions of this theory of moral status, an individual's right to life would wax and wane in correspondence with the quality of impairment occasioned by his or her state of delirium. Arguably, there could be periods over the course of the day during which it would be morally permissible to kill a patient so afflicted on the grounds that there were windows of time when the individual was not truly self-conscious.

Potential for Self-Consciousness

Perhaps in recognition of the serious flaws attending the self-consciousness theory, some have suggested that moral status ought to be based upon the mere *potential* for self-consciousness. Indeed, this view provides a remedy for

10. American Psychiatric Association, *Quick Reference to the Diagnostic Criteria From DSM-IV* (Washington, D.C.: American Psychiatric Association, 1994), pp. 81-82.

many of the shortcomings of the self-consciousness theory. Since the new-born, comatose, and delirious all possess the potential for self-consciousness, the potentiality view prohibits individuals from being killed simply because they lack self-consciousness at a given moment. Philip E. Devine summarizes the potentiality theory as follows:

> According to this principle, there is a property, self-consciousness or the use of speech for instance, such that (i) it is possessed by adult humans, (ii) it endows any organism possessing it with a serious right to life, and (iii) it is such that any organism potentially possessing it has a serious right to life even now — where an organism possesses a property potentially if it will come to have that property under normal conditions for development.[11]

Although the potentiality theory appears to represent an improvement over its predecessor, it retains some problematic features. One critique of the potentiality argument has been advanced by H. Tristram Engelhardt Jr. It points out a basic error in logic. In summary, Engelhardt's response to the potentiality theory states that, inasmuch as X is merely a potential Y, then it stands to reason that X is not Y.[12] In the context of the present discussion, if an embryo is merely a potential person, then an embryo is not a person.

This objection appears to have some validity for at least one form of the potentiality theory. Here, semantics are important, and Stephen Buckle reminds us that potentiality may be argued from two different and distinct positions: the potential to *become* and the potential to *produce*.[13] When X is *becoming* Y, there is a continuity of identity between X and Y, and, as such, the former is afforded the rights and privileges of the latter. Under this construction, the developing embryo and fetus would be extended the rights and privileges of a human adult, as the former *will be* the latter. However, if X is considered to be but a necessary component in *producing* Y, then a continuity of identity is not maintained between the two and the rights and privileges of the latter are not transferred to the former. The former will never be the latter but will only be *part of* the latter.

11. Philip E. Devine, *The Ethics of Homicide* (Ithaca, N.Y.: Cornell University Press, 1978), pp. 94-95.

12. H. Tristram Engelhardt Jr., *The Foundations of Bioethics* (New York: Oxford University Press, 1986), p. 111.

13. Stephen Buckle, "Arguing from Potential," *Bioethics* 2 (1988): 227-53. Carson Strong does a fine job of unpacking the implications of these two senses of the term "potentiality," and the reader is referred to his treatment of the subject. See Strong, *Ethics in Reproductive and Perinatal Medicine*, pp. 45-46.

Engelhardt's argument highlights the difficulties that attach to this latter sense of the potentiality theory, namely, *producing*. Indeed, if X merely has the potential of producing Y, then X is not Y; Y does not maintain an identity with X and Y is an entirely different entity. However, when X is in the process of becoming Y, Y maintains an identity with X. The difference between X and Y is not one of quality, but of time. X is Y, just at an earlier stage of development.

This distinction helps to explain why ovum and semen do not have the same moral status as a fertilized egg. Whereas the ovum and semen have the potential of *producing* a human being (should a sperm penetrate and begin fertilizing the ovum), a fertilized ovum is *becoming* a human being. This distinction permits the destruction of ovum and sperm, but not the fertilized ovum. The fertilized ovum maintains continuity of identity with the human adult that it is becoming. Therefore, the former ought to be assigned with the same moral standing and right to life as the latter.

Finally, Devine states that "an organism possesses a property potentially if it will come to have that property under normal conditions for development."[14] Should "normal conditions for development" be provided from the moment of conception, an organism will eventually come to have self-consciousness. Accordingly, Devine's position supports acknowledging a human being's right to life from the moment of conception.

Sentience

Yet another proposal for moral status has been advanced by L. W. Sumner who has suggested that sentience should form the basis for a right to life. Strong has defined sentience as "the capacity for feeling or perceiving."[15] Sumner maintains that moral status and, hence, a right to life, necessarily accompanies the ability to perceive pain. Furthermore, Sumner believes that sentience is acquired at some time during the second trimester and that a fetus should be afforded moral standing late in the second trimester or early in the third trimester.[16] Embryos, then, would not have a right to life.

This theory has a number of serious weaknesses. For example, Strong notes that if a right to life is based upon sentience, then adult non-human an-

14. Devine, *Ethics of Homicide*, p. 95.
15. Strong, *Ethics in Reproductive and Perinatal Medicine*, p. 48.
16. L. W. Sumner, *Abortion and Moral Theory* (Princeton: Princeton University Press, 1981), p. 152.

imals that have a similar degree of sentience to that of late-second-trimester human fetuses should be regarded as possessing a similar right to life.[17]

Furthermore, it is not at all clear that sentience is acquired only late in the second trimester. Indeed, Sumner's contention is highly speculative. For instance, a developing fetus may begin to exhibit reflex responses quite early in its development — as early as forty to forty-two days post-conception.[18] Moreover, it may be that the early developing embryo may be sentient prior to its ability to exhibit reflex responses to stimuli, due to the immaturity of its nervous system.

The theory of sentience is also marred by its arbitrariness. The developing fetus triggers labor and is usually born between 255 and 275 days after conception. This places the end of the second trimester at some time between 170 and 183 days (nearly a two-week window of time). If a right to life is acquired "some time late in the second trimester," it begs the question of precisely how much latitude ought to be permitted either side of the 170-183 day time frame. Are 169 and 184 days still considered "late in the second trimester"? How about 165 and 188 days? Surely if one does not have a right to life at 189 days, one more day cannot make that much difference. So, would it be morally permissible to abort a 190 day fetus? Just how broad is the "late second trimester"? In short, the theory of sentience is pragmatically as well as theoretically flawed and, therefore, fails to provide an adequate foundation for the moral status of the human embryo.

Viability

Yet another approach states that the unborn individual acquires a right to life at the moment that it reaches a stage of sufficient maturation so as to allow it to survive outside of the mother's womb, albeit with some artificial means of assistance. This theory of moral status is reflected in the 1973 landmark U.S. Supreme Court decision known as *Roe v. Wade,* in which the Court ruled that a woman's right to secure an abortion is but one aspect of the more general-

17. Strong, *Ethics in Reproductive and Perinatal Medicine,* pp. 48-49.

18. Chronologies of human embryonic development are widely available. For example, see N. Okado, S. Kakimi, and T. Kojima, "Synaptogenesis in the Cervical Cord of the Human Embryo: Sequence of Synapse Formation in a Spinal Reflex Pathway," *Journal of Comparative Neurology* 3 (April 1979): 491-518; Nigel M. de S. Cameron and Pamela F. Simms, *Abortion: The Crisis in Morals and Medicine* (Leicester, England: InterVarsity, 1986), pp. 71-84; and John S. Feinberg and Paul D. Feinberg, *Ethics for a Brave New World* (Wheaton, Ill.: Crossway Books, 1993), pp. 53-55.

ized right of privacy.[19] However, the Supreme Court noted that the U.S. Constitution makes no explicit reference to a purported right of privacy and that, even if implicitly rooted in the Constitution, such a "right" is not absolute.

Despite the prominence of this court decision within American culture, the arguments of *Roe v. Wade* are poorly understood, frequently misrepresented, and often inappropriately applied. To hear many advocates of abortion argue, one would come to think that *Roe v. Wade* makes constitutional provision for a woman to terminate her pregnancy in a manner, at a time, and for reasons of her sole choosing. Indeed, our public discourse is now replete with slogans such as "pro-choice" and "abortion-on-demand." However, such notions are, at the very least, potentially misleading if not patently false. According to the Supreme Court in *Roe v. Wade:* "some . . . argue that the woman's right is absolute and that she is entitled to terminate her pregnancy at whatever time, in whatever way, and for whatever reason she alone chooses. With this we do not agree."[20]

Rather, the Supreme Court stated that viability is the point at which the "state interests as to protection of health, medical standards, and prenatal life, become dominant."[21] Additionally, the Court held that, "With respect to the State's important and legitimate interest in potential life, the 'compelling' point is at viability."[22] Therefore, viability serves as the dividing line between a woman's right to self-determination and the state's "compelling" interest in protecting the unborn individual's right to life.[23]

---

19. *Roe v. Wade,* 410 U.S. 113 (1973).

20. Ibid., p. 153.

21. Ibid., p. 155.

22. Ibid., p. 163.

23. However, it should be kept in mind that the state's "compelling interest" is trumped in practice by considerations for the "health" of the mother. In discussing those factors that comprise "health," the Court stated that *Roe v. Wade* must be understood in light of *Doe v. Bolton,* 410 U.S. 179 (1973). The Court concluded that in determining the "health" of the mother, all factors must be considered including, physical, emotional, financial, age, the stress of having an "unwanted child," the number of children the mother already has, and the stigma of unwed motherhood in society (*Doe v. Bolton,* 410 U.S. at 191-97, and *Roe v. Wade,* 410 U.S. at 153). So, on the one hand, the Court stated that a woman's right to an abortion is not absolute and that she is not entitled to terminate her pregnancy at whatever time, in whatever way, and for whatever reason she alone chooses. On the other hand, the subjective and sweeping considerations allowed by the Court virtually ensure that a woman can secure an abortion on demand, thereby contradicting the Court's own stated position. See Dennis J. Horan and Thomas J. Balch, "Roe v. Wade: No Justification in History, Law; or Logic," in *Abortion and the Constitution: Reversing* Roe v. Wade *Through the Courts,* ed. Dennis J. Horan et al. (Washington, D.C.: Georgetown University Press, 1987), p. 59.

The theory that a right to life is acquired once a fetus attains viability faces several serious difficulties. First, Frost, Chudwin, and Wikler call into question the logical progression of thought in the viability argument: "Why should a fetus's capacity to live independently be a reason to forbid the mother from forcing it to live independently?"[24] Moreover, it is not at all clear why the fact that an entity cannot live independently outside of a specified environment provides adequate moral grounding for permission to remove the entity from such an environment.

An additional objection to the viability theory is rooted in today's rapid advancements in technology and medicine. Specifically, the Supreme Court in *Roe v. Wade* stated that "Viability is usually placed at about seven months (28 weeks) but may occur earlier, even at 24 weeks."[25] However, one must bear in mind that the Court's statement on viability reflected the prevailing medical wisdom of its day.[26] Modern technological and medical procedures continue to push ever earlier the point at which life can be supported outside of the womb.[27] Should future advancements include the ability to artificially sustain life outside of the womb from its inception, then viability would begin at the very moment of fertilization.[28] This would result in an assignment to the earliest embryo the same moral status as that of a normal adult human being, along with a corresponding right to life. In other words, the problem with viability as the basis for the moral status of the embryo is that it is a mea-

24. Norman Frost, David Chudwin, and Daniel Wikler, "The Limited Moral Significance of Fetal Viability," *Hastings Center Report* 10 (1980): 10-13, at 13.

25. *Roe v. Wade*, p. 160.

26. In rendering its 1973 opinion regarding the point of fetal viability, the Supreme Court cited two sources in footnotes. These were: L. Hellman and J. Pritchard, *Williams Obstetrics*, 14th ed. (1971), p. 493; and *Dorland's Illustrated Medical Dictionary*, 24th ed. (1965), p. 1689. Accordingly, it appears that the Court relied on information reflecting the state of medical knowledge between 1965 and 1971.

27. Carson Strong suggests that the current gestational age at which point a fetus attains viability is approximately twenty-two to twenty-four weeks. Strong, *Ethics in Reproductive and Perinatal Medicine*, p. 53.

28. The Reuters news service reported on 18 July 1997 that Japanese scientist and gynecologist Yoshinori Kuwabara has developed an artificial womb that is capable of sustaining goat fetuses for a period of up to three weeks. The artificial womb consists of a clear rectangular plastic casing filled with amniotic fluid that is maintained at body temperature. Several devices necessary to support vital functions, including a dialysis machine that oxygenates and cleans the fetus's blood, are connected to the umbilical cord. Dr. Kuwabara stated that the aim of the project is to eventually use the technology to sustain human fetuses outside the womb, "but it will take maybe 10 years." The article, entitled "Japanese Scientist Develops Artificial Womb," may be accessed via the Internet at: *www.wels.net/wlfl/country/japan97.html*.

sure of the technology available rather than a measure of the human entity involved.

Yet another hurdle faced by the proponents of the viability theory of moral status has been recently erected by the California Supreme Court.[29] In its 1994 decision in the matter of *People v. Davis*,[30] the court held that the unborn ought to be afforded the rights and protections provided under the California murder statute: "Murder is the unlawful killing of a human being, or a fetus, with malice aforethought."[31] The Court defined a "fetus" as "the unborn offspring in the postembryonic period . . . [which begins] seven or eight weeks after fertilization."[32] The California Supreme Court's decision makes it a capital offense, punishable by death, to cause the death of an unborn child who is as young as seven to eight weeks gestational age.

California also has a statute to protect the rights of the unborn in the event of prenatal personal injury (among other legal rights afforded to the unborn). This statute affirms that "a child conceived, but not yet born, is deemed an existing person, so far as necessary for the child's interests in the event of the child's subsequent birth."[33] In its 1997 decision in the matter of *Snyder v. Michael's Stores*, the California Supreme Court extended the rights addressed in this statute to those injured as a result of a worker's compensation injury.[34] In this case, a child's injuries were caused in utero by the mother's exposure to carbon monoxide fumes while at work. The court found that the little girl, even in utero, was a distinct and separate person and not simply a biological extension of her mother.

It is simply inconsistent to regard a fetus of seven or eight weeks as a distinct and separate person from its mother in matters of murder or personal injury, but not in matters of abortion. Put another way, terminating a pregnancy by abortion cannot be considered a "constitutional right," while murder or injury of the fetus by another is a crime or civil violation. Although abortion and murder are different in some respects (most significantly in terms of the mother's wishes), each of these acts is now predicated upon inconsistent and mutually exclusive views of when human life begins. It

---

29. I gratefully acknowledge the assistance of Kamrin J. Evans, J.D., in helping me to locate the legal cases *People v. Davis* and *Snyder v. Michael's Stores*. She has also written on the meaning and significance of these cases in her article, "Public Policy Report: Is an Unborn Child a Person?" *Veritas Christian Research Ministries Newsletter* 1 (summer 1998): 2, 4.

30. *People v. Davis*, 872 P.2d 591 (Cal. 1994).

31. Calif. Penal Code §187 (a).

32. *People v. Davis*, p. 599.

33. Calif. Civil Code §43.

34. *Snyder v. Michael's Stores*, 945 P.2d 781 (Cal. 1997).

will likely become increasingly difficult for abortion advocates to argue that the developing fetus does not have a right to life until it attains viability while other segments of society are quickly beginning to view the developing fetus as a separate and distinct person who is entitled to the same rights, protections, and privileges of adult human beings.

Logical, technological, and legal problems render the viability theory of moral status seriously deficient. Moreover, while it might seem to deny that an embryo has a right to life, as soon as an artificial womb is developed the viability theory will explicitly support this right.

Similarity

Carson Strong advances his own theory concerning the moral status of the unborn. He suggests that a right to life is incrementally conferred upon the developing embryo/fetus as it progressively attains an overall degree of physical similarity to "normal adult human beings."[35] Strong believes that physical resemblance is a morally relevant factor in assigning a right to life because, "psychologically speaking, we are more likely mentally to associate paradigmatic persons with individuals who *look like* the paradigm than we are to associate them with individuals who do not look like the paradigm."[36] In addition to physical resemblance, Strong adds that potentiality for self-consciousness, sentience, viability, and birth are, likewise, "morally relevant similarities" to those possessed by "normal adult human beings."[37]

While acknowledging that each of these characteristics — in and of themselves — is insufficient to ground the moral status of the human embryo and fetus, Strong believes that the aggregate combination of these similarities to the normal human adult is "significant enough to warrant conferring upon infants serious moral interest, including a right to life."[38] As one moves progressively backward on the developmental continuum from birth to conception (i.e., infant, advanced fetus, intermediate fetus, early fetus, embryo, and early embryo), the entity is afforded a progressively lower moral status. For example, in Strong's words: "We might say that advanced fetuses should have a conferred right to life, but one that is not quite as strong as that of infants."[39]

35. Strong, *Ethics in Reproductive and Perinatal Medicine*, pp. 56-60.
36. Ibid., p. 57. Italics in original.
37. Ibid., pp. 57-58.
38. Ibid., p. 58.
39. Ibid., p. 59.

The theory of similarity, though provocative, suffers from many serious flaws, some of which can be summarized here. First, the theory cannot explain at what point the developing fetus acquires a right to life. Rather, it is content to suggest that from the moment of conception to early infancy, an entity is progressively acquiring a higher moral status. Such an understanding may be subject to many interpretations — none being final.

Second, what constitutes a "normal" human adult? Is physical similarity enough? How would moral standing be conferred in cases of physical deformity? Do human adults who lack certain physical features possess relatively lower moral standing and, hence, less of a right to life? How is a physical interpretation of "normalcy" to be tolerated in a society that simultaneously upholds the principles of the Americans with Disabilities Act? This act reminds us that *all* human beings possess rights, dignity, value, and worth, regardless of any perceived physical limitation.

Third, who will decide what "looks like" a normal human adult? One person may see the requisite "physical resemblance" at twenty-eight weeks. Someone else may see compelling similarity at eighteen weeks. An individual desirous of securing a late-term abortion may merely claim that she (or he) is unable to see the same degree of physical resemblance as another person.

Finally, the various forms of similarity cited by Strong are each riddled with difficulties. For instance, Strong claims that "sentience is the basis of moral interests. . . ."[40] However, such is merely an assertion. Without marshaling support for his claim and demonstrating the relevance of any support to the claim advanced, his contention fails to rise to the level of an argument. Problems with appeals to sentience and other such attributes as the basis for moral standing have been discussed above.

Scriptural Theories on the Moral Status of the Embryo

Having considered several of the more common non-scriptural theories concerning the moral standing of the embryo and fetus, we now turn to a discussion of scripturally based theories. From the Bible, both birth and conception have been frequently argued as the event at which time an entity has a right to life. Here, both will be considered in turn.

40. Ibid.

Birth

One argument that may be advanced from the Scriptures is that a right to life is acquired at the moment of birth. According to this view, embryos would not have such a right. Genesis 2:7 states that "the Lord God formed man out of the clay of the ground and blew into his nostrils the breath of life, and so man became a living being." According to this theory, as Adam became "a living being" at the moment he received breath, so all humans gain a right to life at the moment they take their first breath — namely, at birth.

There are several significant difficulties with this theory of moral status. First, this view stands only if one is willing to argue that the various factors involved in the creation of Adam ought to be considered normative. However, such an approach is highly problematic. In the creation narrative, God personally fashions Adam out of the clay of the ground — we are born of a mother's womb. God created Adam as a full human adult — we begin with conception, proceed through gestation, and are born. The first oxygen that Adam received was expelled from the very mouth of God — we first receive oxygen while still in the womb. God created Adam as *néphesh,* or soul. He was a complete adult — we proceed through various stages of maturation.[41] Indeed, there are too many unique features involved in the creation of Adam for one to reasonably argue that the various characteristics attending to the beginning of his life should be considered normative.

Second, when the biblical genealogies record that "so-and-so begat so-and-so," no precise differentiation between conception and birth is intended. Rather, it appears that God has crowned human beings with a spiritual-moral nature and that all requisite minimal features of God's image are present in each stage of humanity between conception and birth.[42] In support of this argument, one need only consider the poem of David, to be sung by all (Ps. 51), and note that he considered himself to be a spiritual being between the time of conception and birth (Ps. 51:7-8). David remarked that at the very moment of conception he was both sinful and in possession of a conscience. Elsewhere, David states that God had been his guide from the moment that he was formed and that he later came to trust God at his mother's breast (Ps. 22:10). Furthermore, Job provides us with a very transparent metaphor in which he views himself as having been formed and fashioned in the womb by the hands of God.

41. Johannes Pedersen, *Israel, Its Life and Culture I-II* (London: Oxford University Press, 1926), pp. 99-181.
42. Bruce K. Waltke, "Reflections from the Old Testament on Abortion," reprinted in *Journal of the Christian Medical Society* 19/1 (1988): 24-28.

> Your hands have formed me and fashioned me; will you then turn and destroy me? Oh, remember that you fashioned me from clay! Will you then bring me down to dust again? Did you not pour me out like milk, and thicken me like cheese? With skin and flesh you clothed me, with bones and sinews knit me together. Grace and favor you granted me, and your providence has preserved my spirit. (Job 10:8-12)

So while Adam was personally formed by God outside of the womb, here we find that humans are personally formed by God inside the womb. Everything of significance is formed long before breath comes into the picture. In fact, as explained below, science now tells us that the definitive "knitting together" is the construction of the individual's unique genetic code.

The theory of basing moral standing on birth or breath, then, fails on several grounds. It is inadequate exegetically, theologically, and scientifically.

Conception

The other prominent scriptural theory concerning moral status regards conception as the defining event. According to this theory, an embryo has the full moral status of a human being, complete with a right to life, from the moment of conception.

One of the more intriguing arguments concerning the conception theory has been advanced by Nigel Cameron.[43] Cameron bases his argument on the theological implications of the incarnation of Jesus Christ. According to this view, the life of Jesus Christ must have begun in his embryonic biological state due to the fact that this is the time at which he was conceived (rather than being created as a fully functioning person) and that such was necessary in order for him to take on true humanity. Accordingly, it was at the moment of conception that Christ's life story began.

The reason God could take up residency within a human being is that humans are already created in the image of God. Therefore, in order for God to become truly a human being, his image must be present in the embryo prior to any quality of functioning on the part of the embryo. In other words, Mary's "fertilized ovum" (or first embryonic cell — however God "knit it together") must have contained all of the necessary qualities of God's image in it in order for God to assume the beginning of his earthly life in its substance. As Cameron summarizes, "for God to become man by the miraculous fertil-

---

43. Nigel M. de S. Cameron, ed., *Embryos and Ethics* (Edinburgh: Rutherford House Books, 1987), pp. 12-13.

izing of one of Mary's ova, it is necessary that a fertilized ovum should be im-age-bearing already and of its own nature."[44]

One imprecision in Cameron's theory arises from his use of the term "fertilized."[45] Modern biology holds that fertilization is a process over time, not an event in time.[46] For some, this process includes the period of time during which the sperm is being prepared and then is active in the male and female reproductive tracts prior to fusing with the egg. Many more would at least include the nearly twenty-four-hour period after the sperm penetrates the outer surface of an egg but before the chromosomes of both the sperm and egg align themselves and form a new genome.[47] The aligning process, called "syngamy," culminates in the fertilized egg becoming a single-cell zygote. Accordingly, conception can be defined as "the formation of a viable zygote."[48] Traditionally the term "embryo" has not been employed until the zygote completes about two weeks of development.[49] However, more recently the term has come to be used from the time of conception onward — a practice that is followed in this chapter.

In implying that the process of fertilization must be a completed act (i.e., "fertilized ovum") prior to locating the image of God within its substance, Cameron leaves an unanswered question. What is the moral status of the entity during the early and intermediate stages of fertilization? Would it be morally permissible to terminate the process of fertilization at some point prior to the twenty-fourth hour? The answers to these ques-

44. Ibid.

45. I wish here to express my appreciation to Vincent Ling, Ph.D., staff scientist in immunology at the Genetics Institute in Cambridge, Mass. Dr. Ling reviewed and commented on the following section of the manuscript, which deals with human biology and embryology. I gratefully acknowledge his assistance.

46. For a more complete description of the fertilization process, see C. R. Austin, *Human Embryos* (Oxford: Oxford University Press, 1989), pp. 1-41; and Clifford Grobstein, "The Early Development of Human Embryos," *Journal of Medicine and Philosophy* 10 (1985): 213-36.

47. W. Richard Dukelow and Kyle B. Dukelow, "Fertilization," in *The Encyclopedia of Human Biology*, vol. 3, ed. Renato Dulbecco (San Diego, Calif.: Academic Press, 1991), p. 599.

48. *Webster's Third New International Dictionary* (Springfield, Mass.: G. & C. Merriam, 1981), p. 469.

49. At about the fourteenth day, a dark-colored linear streak appears along the axis of the fertilized egg. At the fore-end of this streak — known as the "primitive streak" — the head will eventually be formed. The appearance of the primitive streak marks the beginning of what has traditionally been called the embryonic stage. See Grobstein, "Early Development of Human Embryos"; and Christopher Vaughan, *How Life Begins: The Science of Life in the Womb* (New York: Times Books, 1996), p. 8.

tions appear to have considerable bearing upon the moral appropriateness of certain contraceptive and birth control methods. By stating that the "fertilized ovum" is image-bearing, Cameron (perhaps quite unwittingly) has left open the possibility that it is morally permissible to terminate an ovum that is still in the process of fertilization, and not yet completely fertilized. A stronger alternative would be to hold that the image of God is located in the "fertilizing ovum," rather than in the "fertilized ovum." Such would place the imputation of God's image in human beings at the initial moment of conception (i.e., fusion of sperm and egg), rather than at the end of the fertilization process. The entire, unique genetic code is present from the point of initial fusion, even though repositioning within the cell will still take place.

The above refinement, however, does not resolve all of the issues of concern. Clinically, it is exceedingly difficult to determine the precise moment at which conception begins because women can vary considerably in their ovulation cycles. Often, women are instructed to note the date of the beginning of their last menstrual period and count that day as zero, or the time of conception. However, most women produce an egg fourteen days after their last menstrual period — resulting in the embryo already being "two weeks old" at the moment of conception.[50] The actual age of the embryo then continues to be approximately two weeks off throughout the course of the pregnancy. Even in situations where a woman is able to identify the last occurrence of sexual intercourse, it remains nearly impossible to state with any degree of certainty the precise moment of conception.

In many cases, this point is moot. The embryo begins implantation into the uterine wall at approximately six to nine days and is usually completely imbedded into the uterine wall at ten days. Once implanted, pregnancy is regarded as established. Therefore, by the time a woman has learned that she is pregnant, fertilization has already completely taken place.

In the biblical narrative, the angel Gabriel advises the virgin Mary that she will "conceive and bear a son" (Luke 1:31). Emphasis is placed upon the very beginning point at which her son's life will begin. In a similar way it is most plausible to hold that the image of God is imparted to human beings at the beginning of the process of fertilization, and that the right to life is, therefore, conferred at the initial moment of conception.[51] On such a view, any

---

50. Vaughan, *How Life Begins*, p. 8.

51. The purpose of this analysis is to determine whether or not the human embryo has a right to life, not to define personhood. Whether a person is dying every time an embryo perishes must be the subject of another discussion.

abortifacients or procedures that cause the destruction of human life after this moment would be morally prohibited.

*   *   *

The purpose of this discussion has been to examine several of the more prominent secular and religious theories concerning the moral status of the human embryo. The results of this study suggest that a right to life is conferred simultaneously with the imputation of the *imago Dei* (image of God) at the earliest moment of conception. Such a view prohibits moral status from being attached to, or predicated upon, any self-directed or purposeful behavior, attitude or sentiment. Rather, where human biological life has begun (regardless of its limitations or stage of development), there is God's image. Although each life is differentiated in many ways from the initial moment of conception, even human embryos possess the very nature, value, and dignity of being the *imago Dei*. As such, they ought to be regarded as having a right to life.

PART II

SPECIFIC TECHNOLOGIES

CHAPTER 5

Sexual Ethics and Reproductive Technologies

Dennis P. Hollinger, Ph.D.

The revolution in reproduction is not merely a technological achievement. It invariably raises ethical issues. There is a tendency today to divorce technology from any moral scrutiny, and this is particularly true in the case of reproductive technologies. But these technologies are not neutral, for their very nature raises theological and ethical questions.

Reproductive technologies can, of course, be analyzed from various ethical viewpoints. For some, the overriding issue will be the principle of compassion for couples who face the potential reality of childlessness. For others the issue will be nature and the extent to which the various technologies work with nature or act contrary to its givens. For still others the primary motif for ethical reflection will be the nature of personhood, and the effect of a given technology upon human dignity and the nature of humanity. These and other approaches all have their place in an ethical analysis of reproductive technologies.

But clearly one of the issues that we cannot ignore in the midst of this revolution is our understanding of sex. Because the technologies are attempts to assist or even commence the reproductive process, we are forced to reflect on an understanding of reproduction in relation to the way it has always heretofore occurred, namely through sexual intercourse. What is the nature and meaning of sex? It is in answering this question that we come to understand the proper uses of this good gift of God. And it is from our understanding of sex that we also come to see both the potential and limitations of reproductive technologies.

As we probe these issues it is important to keep in mind a salient distinction, that between sex and sexuality. Whereas sex is an activity, sexuality refers to our maleness and femaleness as human beings. It is far broader than acts of sexual intimacy, for it involves our identity. Our sexuality is tied to our very capacity to relate to other people. As Stanley Grenz puts it:

> Sexuality comprises all aspects of the human person that are related to existence as male and female. Our sexuality, therefore, is a powerful, deep, and mysterious aspect of our being. It constitutes a fundamental distinction between the two ways of being human (i.e., as male and female). This distinction plays a crucial role in our development toward becoming individuals and our existence as social beings.[1]

Sexual intimacy is then just one aspect of our sexuality. This essay will focus primarily on sexual intimacy and its meaning in relationship to reproductive technologies.

The Meaning and Purposes of Sex

The first and perhaps most important thing we can say about sex is that it is a good gift of God. In the creation account of Genesis 1, God "created people in his own image, in the image of God he created humanity; male and female he created them. . . . God saw all that he had made, and it was very good" (Gen. 1:27, 31a). From the very beginning Adam and Eve were meant to experience the goodness of God's gift, resulting from their being made male and female. Without reservation they could walk with God in the garden by day and be spontaneously sexually intimate at night.[2]

Christians down through the centuries, however, have not always been at peace with this reality. Early on, the church was plagued by notions of asceticism that disparaged the physical body and its impulses. Concerned about sexual drives, which can be perverted due to our fallenness, church leaders cast doubt on the essenge is that it produces virgins.[3]

The goodness of sexual intimacy, however, does not imply an unbridled use of this gift. If asceticism was a threat to the church's understanding of sex, it is naturalism that has been secular society's greatest challenge. Naturalism

---

1. Stanleytial goodness of this gift from creation. For example, Origen, the great third-century theologian, castrated himself; and Jerome, the translator of the Latin Vulgate in the fourth and fifth centuries, argued that the only good of marria Grenz, *Sexual Ethics: An Evangelical Perspective* (Louisville: Westminster, 1990), p. 21.

rightly affirms the goodness of sex but overlooks the fact that there is inherent meaning to the act and that parameters are given in the midst of its goodness. The naturalistic philosophy sees sex as primarily a natural drive that society chooses to endow with meaning, significance, and limitations. The pleasure dimension is usually primary and often to the exclusion of other dimensions. The naturalistic perspective, so prominent in our own time, is clearly reflected in the writings of the late Michelle Foucault, French philosopher and self-proclaimed pedophile. In his castigation of moral principles, Foucault argues that rules are empty, violent, and used by those in power merely to get their way. He argues, "If repression has indeed been the fundamental link between power, knowledge, and sexuality since the classical age, it stands to reason that we will not be able to free ourselves from it except at considerable cost: nothing less than a transgression of laws, a lifting of prohibitions . . . a reinstating of pleasure within reality, and a whole new economy in the mechanisms of power will be required."[4]

In contrast to both asceticism and naturalism, biblical faith affirms the goodness of sexual intimacy that is in line with its intended purposes. There is inherent meaning to the sexual act. That is affirmed biblically as well through natural observation and historical experience. In particular, God has given the gift of sex to humanity for four main purposes, and a legitimate sexual act must embody or be experienced within the context of these four purposes.

The Consummation of Marriage

Biblically and historically in many societies the sexual act has served as a consummating act to the marriage. Genesis 2:24-25 states: "For this reason a man shall leave his father and mother and be united to his wife, and they will become one flesh. The man and his wife were both naked, and they felt no shame." This is perhaps the closest we come to a definition of marriage in the Bible, and it is one that historically has been experienced around the world. It embodies three dimensions: a change of status ("leaving father and mother"), a commitment ("united to his wife"), and the consummation ("they will become one flesh"). Wherever this later phrase is used in the Bible it is with reference to sexual relations. In sexual intimacy, something profound happens to two human beings, for this is the act that brings the totality of the self into

2. See Lewis Smedes, *Sex for Christians* (Grand Rapids: Eerdmans, 1976), p. 17.
3. *Post-Nicene Fathers*, vol. 6, p. 30. See also pp. 71, 77.

the fullest intimacy with another human being. It is the ultimate act of trust, as husband and wife abandon themselves vulnerably to each other. As Lewis Smedes puts it:

> The physical side of sexual intercourse is a sign of what ought to happen on the inside. It is the final physical intimacy. Two bodies are never closer: penetration has the mystique of union, and the orgasmic finale is the exploding climax of one person's abandonment to another, the most fierce and yet most sensitive experience of trust.[5]

In relationships in general we may share many things with another person that we also share in marriage, such as friendship, deep empathy, material resources, and time. But what sets the marriage relationship apart from all other relationships is the extent of the commitment and the physical union whereby the two become one flesh. Every act of sexual intercourse by husband and wife is a reaffirmation of that new one flesh relationship which was formed at marriage. It solidifies, reaffirms, and nurtures the deep organic bond which is at the heart of marriage. Sexual intimacy completes the marriage act, but with each act the couple reaffirms the uniqueness, mystery, and solidity of their relationship. The continuation of one-flesh intimacy in marriage sets the relationship apart from all others; it is a boundary for the covenant that has been established.

Procreation

From both nature and the Bible it is evident that procreation is an inherent part of sexual relations. After God created humanity in his image, male and female, "God blessed them and said to them, 'be fruitful and increase in number; fill the earth and subdue it'" (Gen. 1:28a). Until recently, sexual relations was the only means by which human life on earth could continue. God's design was that human life begin in the most intimate, loving act that two humans can experience. When humanity and society deviate from that norm the negative consequences for both personal development and societal stability are evident.

There is a tendency today in thought and practice to radically divorce sexual relations and procreation. It occurs when people engage in sexual intimacy with no thought of, or responsibility for, the possibility of new life, and it occurs in some new forms of reproduction. Sex and procreation are pulled

4. Michelle Foucault, *The History of Sexuality: Vol. 1, An Introduction* (New York:

asunder in such a way that many people see them as only tangentially related, and then only as one option.

But every sexual act is by nature procreative. This is not to say that every act will procreate, or should procreate, or can procreate, but rather that it is the act which biblically, historically, and naturally is linked to the conception of new life. A legitimate sexual act is one in which the two people are willing and able to care for the potential fruit that may come from their intimacy, for the nature of that intimacy is a life-producing action. This view does not necessarily imply a rejection of contraception, which can be defended on the grounds of both stewardship and the multiple purposes of sex.[6] But it does mean that every sexual relationship should be entered into with the realization that this is a procreative act and with a willingness to bear the potential fruit that could come from this act. The context in which this is most clearly possible is a marriage between a man and a woman. As the late Paul Ramsey argued:

> The fact that God joined together love and progeny — or the unitive and procreative purposes of sex and marriage — is held in honor, and not torn asunder, even when a couple for grave, or for what in their case is to them sufficient, reason adopt a lifelong policy of planned unparenthood. . . . Such married partners are still saying by their actions that if either had a child this would be from their one-flesh unity with each other and not apart from this.[7]

Just as it is possible to engage in sexual relations without any thought or responsibility for procreation, so it is now possible to procreate without sexual intimacy. In such cases, procreation is pulled apart from the loving, covenantal context that can best nurture the new life. When procreation and sex are pulled apart from each other, it is not nature or moral laws that are the losers; it is rather human beings who lose. To separate what God has intended to be linked distorts their fullest meaning and the deep, inward joy that God's good gifts can bring.

Vintage, 1980), p. 5.

5. Smedes, *Sex for Christians*, p. 132.

6. While such a view differs from the official Roman Catholic position (which argues that the procreative and the unitive dimensions are tied together in such a way that technological contraception is unacceptable), most of what is argued here is in line with classical Catholic — and Protestant — teaching. See, for example, the Vatican's "Instruction of Respect for Human Life in Its Origin and on the Dignity of Procreation," published in *Origins* 16, 40 (19 March 1987). The document is often referred to as *Donum Vitae*.

Love

By nature we know that sex is linked to love. When a man and woman love each other and commit to each other, there is a natural drive to express that love in physical intimacy. What is known by nature is affirmed in Scripture, the Song of Solomon being perhaps the best example. When asceticism has thwarted the church's appreciation of sex as an inherently good gift of God, this book of the Bible has often been allegorized into the love of God for his people and their love for him. But there is no evidence for such an interpretation from either the book itself or other parts of the Bible. The vivid descriptions of the speaker's lover and the powerful portrayal of emotional appeal all point to the physical expression of love.

> Like an apple tree among the trees of the forest is my lover among the young men. I delight to sit in his shade, and his fruit is sweet to my taste. He has taken me to the banquet hall, and his banner over me is love. (Song of Solomon 2:3-4)

Sexual intimacy is the most intimate, physical, and expressive way of saying, "I love you." It is, of course, not the only way and is hypocritical if not accompanied by the myriad, and often rather mundane, ways of expressing love. But the sexual bond in marriage is both an expression of love and a nurturant of that love. Thus, the sexual dimension tends to wane when the marriage relationship itself has gone sour. Admittedly, not every sexual act within a marriage is filled with deeply emotional love; after all, love is far more than emotion. But the sexual intimacy is within the context of a loving committed relationship, and thus expresses love even when at a particular point in time the emotional impulses may not be as strong as at other times.

Today sex has often moved outside of love to become mere recreation or physical release. For some it is a substitute for real love, while for others sex is merely expressing a physical drive, often to a stranger with whom there is not even an attachment of any sort. Such expressions distort the meaning and the real joy of sex, for the intimacy is not accompanied by the deep inner impulses and commitments that make the act a tender, sensitive giving of self to another. The very language "making love" totally distorts the love dimension in sex, for physical intimacy is an act of giving to and receiving from one with whom we not only feel love, but one we commit to love, as long as we both shall live.

God's design is that sex and procreation be linked to the bond of love. When human life begins and is nurtured in a bond of committed love, it is the best possible context for human development, and the best insurance against the many personal and societal ills that now confront us.

Pleasure

There is indeed another purpose for the good gift of sexual intimacy — pleasure. God gives good gifts for human enjoyment, and sex is one of those gifts, as long as it is not divorced from the other purposes. There is a Christian form of hedonism that gladly accepts pleasure in its proper place as a gift from the hand of God to bring joy in life. As C. S. Lewis once quipped, pleasure is not the invention of the devil. The Song of Solomon extols not only human physical love, but pleasure. Similarly, Proverbs 5:18-19 celebrates the pleasure dimension in the context of an admonition against adultery:

> May your fountain be blessed, and may you rejoice in the wife of your youth. A loving doe, a graceful deer — may her breasts satisfy you always, may you ever be captivated by her love.

The pleasure dimension is obvious by nature, for physical pleasure is an accompanying part of the sexual act. Certain parts of the physical body serve no other function than pleasure in sexual intimacy — for the female the clitoris and for the male the glans penis. They serve no other biological functions. If God made us in this fashion it is certainly an affirmation of pleasure in sexual relations.

When, however, pleasure is divorced from the other purposes of sex, even the pleasure dimension itself suffers. This is because sexual pleasure is far more than genital arousal; it is far deeper and more holistic in nature. The pleasure is physical, emotional and even spiritual when rightly accepted as a good gift of God for his intended purposes. As Richard Foster has noted, "The problem with the topless bars and the pornographic literature of our day is not that they emphasize sexuality too much but that they do not emphasize it enough. They totally eliminate relationship and restrain sexuality to the narrow confines of the genital. They have made sex trivial."[8]

What then is a legitimate sexual act? It is one experienced in the context of the four purposes of God's good gift to humanity. And those four purposes are only experienced in one place, the marriage of a man and a woman. When any one purpose is pulled out of the mix it affects the others. In our own time

7. Paul Ramsey, "A Christian Approach to the Question of Sexual Relations Outside of Marriage," *Journal of Religion* 45 (April 1965): 105.

only pleasure seems to be a sure reality in sex, and even that seems to have lost its lure as many today seek for the bizarre to bring sexual fulfillment. In the sexual mores and practices of our time, the issue is not that one or two of these purposes are missing, but rather that all four have been severed from each other. Hence the inherent meaning of sex is lost. Every sexual act, to be morally legitimate, should take place in the context of these four purposes.

But it is not only certain sexual acts that do injustice to the meaning and purposes of sex. When we look at the new reproductive capabilities, we find a similar divorce occurring. The divinely ordained purposes of sex are often pulled apart as new forms of procreation are removed from the other purposes. Our response to the reproductive revolution must not be simplistically one-sided, for the various technologies call for differing moral responses. And our ethical response to these new possibilities is certainly fueled by more than a theology or ethics of sex. But this framework does alert us to one of the major problems with a number of the technologies; and it is to this issue that we now turn.

Applying Sexual Ethics to Reproductive Technologies

What follows is not a full-scale analysis of each of the major technologies, but rather an application of the view of sex just set forth. In particular, I will examine the ways in which the forms of reproduction impinge on the unity of sexual ethics. If God's intention is that a legitimate sexual act is one that incorporates the four purposes of consummation, procreation, love, and pleasure, then it follows that God's intention for procreation is to occur within the context of the same four dimensions. To what degree do the various technologies maintain or pull apart the fourfold sexual intentions? In what ways do they maintain the unity of the inherent meaning of sex or detract from that meaning?

Artificial Insemination

Artificial insemination (or intrauterine insemination as it is now often called) is the simplest and oldest of the reproductive technologies, with the first incidence occurring in 1884. There are essentially two forms of artificial insemination, AID (artificial insemination by donor) and AIH (artificial insemination by husband), and the ethical response to each is quite different.

From the standpoint of the sexual ethics just described, there is virtually no moral problem with AIH. While the conception does not occur through sexual intercourse of the husband and wife, the conception does take place in the context of the one-flesh relationship of marriage in which sexual intercourse is a normal part. As Scott Rae explains, "The teaching of the creation account is that procreation is to take place within the oneness of a total marriage relationship, not necessarily a specific instance of sexual intercourse."[9] Therefore, AIH does not nullify the meaning and purposes of God's good gift of sex or go contrary to his intention for procreation. The new life is conceived and nurtured within the bonds of a one-flesh relationship, accompanied by love and pleasure.

AID is a very different phenomenon. Generally the sperm donor is anonymous and there is no bond between the mother and the donor, and no responsibility on the part of the donor father. AID is utilized in a number of contexts, for example, where there is a sterile husband, husband with a hereditary disease, single woman wanting a child, or a lesbian couple. It may also be employed for eugenic purposes. AID is certainly not adultery in a technical sense, for there is not a physical union between the male and the female. There is, however, an intrusion of a third party into the marital relationship (if there is a married couple involved at all), and there is a genetic coming together of two people without even a relationship, and with the one parent unwilling to bear responsibility for his own offspring.

From the framework set forth here, there is in AID a radical separation of the various dimensions of sex. Procreation occurs outside the context of a one-flesh relationship in which love, pleasure, and an organic union form the setting for the emergence and development of the new life. The child will never have the joy and inner security of knowing that he or she is the result of a loving, committed relationship that was open to the potential fruit from their union. The extent to which the various dimensions of sex and the creation family paradigm are put asunder is well captured by Daniel Blankenhorn in *Fatherless America:* "The sperm father completes his fatherhood prior to the birth of his child. His fatherhood consists entirely of the biological act of ejaculation. He spreads his seed, nothing more. He is a minimalist father, a one-act dad."[10]

---

8. Richard Foster, *Money, Sex, and Power* (San Francisco: Harper & Row, 1985), p. 92.

9. Scott Rae, *Brave New Families: Biblical Ethics and Reproductive Technologies* (Grand Rapids: Baker Books, 1996), p. 28.

In Vitro Fertilization and Its Variants

Since 1978 and the birth of Louise Brown in England, thousands of children have been born by some technology involving the manipulation of eggs as well as sperm. In vitro fertilization was the technique used to produce the test tube baby Brown. This technique involves combining a sperm and egg in the lab, and then after initial cellular division, placing the zygote into the uterus where gestation occurs normally. There are several variants such as ZIFT (zygote intra-fallopian tube transfer) in which the zygote is placed into the fallopian tube rather than the uterus, and GIFT (gamete intra-fallopian transfer) in which the sperm and an unfertilized egg are inserted into the fallopian tube.

There are several ethical issues and some pragmatic concerns that arise in relation to these technologies, but in none of them is there necessarily a separation of procreation from the other dimensions of sex. While the conception does not take place from intercourse itself, as long as the conception sufficiently involves husband and wife, it takes place in the context of a loving, enjoying one-flesh relationship in which attempts to procreate have been unsuccessful. The gametes are from within the marital relationship, which involves all the basic dimensions of marriage and sex possible, with the addition of technological assistance in bringing the egg and sperm together.

The biggest ethical issue with such technologies, with the exception of GIFT, is what to do with the extra fertilized eggs if there are more than optimally needed for transfer to the mother. The potential loss of human life has been one of the biggest moral concerns here, though it is possible to get around this problem by fertilizing only one egg or transferring to the mother all fertilized eggs. This issue does not arise in GIFT since the conception itself takes place within the fallopian tube. Thomas Elkins and Teresa Iglesias explain these and a broader array of concerns in their chapters on reproductive technologies inside and outside of marriage, elsewhere in this volume.

Surrogacy

In simple terms, surrogacy is bearing a baby for another. In its most common form, a woman designated the surrogate agrees to conceive a child for a couple, utilizing sperm from the husband of the couple via artificial insemination. The surrogate then surrenders the child back to the couple after the baby's birth. Other forms can get quite complicated. In one form, the sperm comes from an anonymous donor rather than the husband of the couple desiring the child, so that neither of the socializing parents are genetically related to the child. In another form, the wife of the couple provides the egg, but due to uterine problems cannot carry the child. Hence the sperm and egg are from the couple (so that the child is biologically theirs), but the child comes to term in the womb of a surrogate.

In surrogacy there is a separation of the various dimensions of sex. While there is often a loving couple desiring to care for the child, the methods employed pull procreation apart from the one-flesh relationship. There is an intrusion of a third party into that marriage, and the procreation is greatly distanced from the marital relationship. The biological ties to both parents are present in the last example above, but the mother does not carry the child through pregnancy. Another human has entered into the physical-emotional-spiritual intimacy of the procreation process. So whereas the other forms of surrogacy conflict directly with the ethical framework presented here, this last case is at least ethically problematic and therefore generally to be avoided.

Some may object, "But is this really any different from an adoption? Isn't surrogacy really a form of adoption? And no one objects to adoption." Several responses to this objection are in order. First, in adoption we are responding to a tragic or unfortunate situation with a loving acceptance of a child into our midst. It is responding to a need that has arisen. In so doing, couples are able to fulfill not only the wonderful responsibility of nurturing a child, but also able to legitimately fulfill the innate desire to have a child. Second, in a positive response to adoption we are not giving moral sanction to the situation in which the child was conceived. Often adoption is necessary because procreation has been pulled apart from the other dimensions of sex. To adopt is not to agree that the action necessitating adoption was good, but is to respond to an unfortunate situation in a morally good fashion. To agree to surrogacy, however, is generally to put our ethical approval upon the severing of procreation from the other dimensions of sex. The clearest exception would be to bring to term an embryo that has been abandoned by the parents.

There are other ethical issues that have been raised regarding surrogacy, such as the commercialization of the womb, the negative impact upon the surrogate, and the identity confusion which might occur for the child. But even if these potential outcomes do not occur, or one finds ways to ameliorate them, there is typically in surrogacy a betrayal of the God-ordained intention for procreation and a pulling asunder of the various dimensions of God's good gift to humanity. These matters are further discussed by Scott Rae in his chapter on surrogacy later in this volume.

Cloning

In cloning we face the most radical separation of sex, procreation, and love. A child would be born not only without sexual intimacy, but for the first time without the "stuff" of sex, namely egg and sperm. In cloning, procreation occurs through removing the nucleus from an adult female egg and replacing it with the genetic material of any adult cell. The egg begins to act as if impregnated and cellular cleavage begins, producing a child who is genetically identical to the adult who provided the genetic material.

The Huxleyan scenarios potentially resulting from cloning have assuredly been overstated, and scientific misinformation has abounded. Clearly one of the major objections to cloning is in its intentionalities, most of which would use the offspring as a means to an end, whether that end be organs, other bodily materials, scientific knowledge, eugenics, or replacing a lost child. But beyond this objection is the problem of radically separating procreation from sex and its multiple purposes. John Kilner discusses this matter and many other concerns related to human cloning in his chapter later in this volume.

What if the egg and genetic material came from a one-flesh, husband-wife relationship? Would this not then be in the context of a procreative, marital bond and thus meet our criteria? In one sense it would, though it is clearly bypassing the very nature of procreation as God has ordained it, namely the fruit of the husband and wife. For in cloning, the genetic link is to only one of the parents. It is not truly the fruit of the loving, enjoying, procreative one-flesh relationship, for one person does not share in the biological connection.

However, there is also something else that is lost in the cloning process, according to a Christian understanding of sex and procreation: the great sense of mystery and a willingness to accept the unknown mystery in procreation. In cloning we totally control the outcome from the start, at least from a genetic perspective. We have determined ahead of time the sex of the child and the genetic makeup of the child, though certainly that genetic makeup will interact with environment in significant ways. In all other forms of reproduction there is mystery and an openness to whatever comes from the fruit of our love. Of course, we know all too well that many do not want to accept that potential fruit and the mystery surrounding it, and hence the high incidence of abortion in our modern world. But in cloning we intentionally bypass the process of mystery with a full, willful determination of the child's biological identity. As Allen Verhey has so eloquently put it, we must ask "whether parenting is properly considered making children to match a specific design, as is clearly the case with cloning, or whether parenting is properly regarded as a disposition to be hospitable to children as given."[11]

*   *   *

In the postmodern world of thought in which we find ourselves, there is a tendency to approach issues like reproductive technologies through the lenses of tolerance, compassion, and personal (or couple) meaning. The notion that there are divine givens in nature to which we ought to conform is a foreign way of thinking in our age. There are in the contemporary frameworks no inherent givens within family, sex, or even human nature other than perhaps self-fulfillment. But a Christian worldview operates from a different paradigm, one that holds that meaning and the nature of things like sex, marriage, and family are given. When we live contrary to those givens and God-ordained intentions, we as human beings are the losers.

Our society is in a crisis over the issue of sex, as contradictions abound on all fronts. Therefore, the church must be clear in its own thinking and actions, as well as in its witness to the world of a better way. Sex is a wonderful and good gift of God. But its goodness is only experienced to the full when the four purposes are operative: consummation, procreation, love, and pleasure. Amidst the reproduction revolution of our time, it is imperative that we hold those dimensions together — for the glory of God, for the good of society, and for true inner fulfillment in our own lives.

10. Daniel Blankenhorn, *Fatherless America: Confronting Our Most Urgent Social Problem* (New York: Basic Books, 1995), p. 171.

CHAPTER 6

Using Donor Eggs and Sperm: The Tragic Face of Anonymity

Teresa Iglesias, D.Phil.

The use of donated gametes (egg and sperm) to conceive a child is skyrocketing — raising profound questions in the process. What does it mean to be human? What does it mean to be gendered (male and female) and to engender — to bring forth new life? What are we and who are we? Indeed these are the basic questions at the center of our concern with modern technological culture, from which asexual forms of human generation have emerged.[1] These questions are addressed personally to each one of us. We are called to respond to them.

The Essence of Existence

The millions of people today who live according to biblical revelation can find answers to these questions about our humanity by turning to a source of fundamental beginnings: the first biblical book. The Jewish community calls this book by its opening words: "In the beginning." Commonly, Christians and others call it Genesis: "Coming to be." Our very first origins are there described in these terms: "In the beginning God created the heavens and the earth" (Gen. 1). The material cosmos, nature, the order of matter from which

---

1. This large question of "what and who we are" has been masterfully dealt with in the context of technological generation, particularly related to cloning, by Leon R. Kass in "The Wisdom of Repugnance," *The New Republic*, 2 June 1997, pp. 17-26.

we come to be, in which we exist and have our being, is God's creation. This is the ground on which we stand. It is not of our making.

John the Evangelist, as a Jewish writer, understood well the significance of the first "In the beginning." He opens his Gospel as a new account of creation with the same words, "In the beginning." In this second version of creation the incarnation of God's Son takes center stage. Jesus is both the eternal Word of God and the "Word made flesh" (John 1:4), a human being, a human body, human matter. In this view of creation, God is not only the creator of matter and its order, but, in the person of his Son, he enters into this order and becomes matter himself, that is, flesh, just like ourselves. And out of incomprehensible graciousness, he is to remain in bodily form, like one of us, for eternity. With Christ our own flesh, our own body, become of divine significance. By the power of the Spirit our flesh becomes also God's dwelling place.

At the center of creation is Jesus. He is the key to its interpretation and understanding. Jesus is also the key to our biblical interpretation. His living example and understanding of biblical revelation is paradigmatic for Christians. We have to trust him as the normative interpretation of God's intention: "This is my own dear Son with whom I am well pleased, listen to him" (Matt. 17:5).

In his own time and social context Jesus was respected for his knowledge of biblical writings. To use our modern terminology, he must have been a thorough student and an assiduous reader of the Scriptures. He knew the book well. He was publicly recognized as a teacher, called "Rabboni" not only by a loving woman, but by experts in the law, the respected lawyers of the time. This recognition is ratified in the questions that these same lawyers address to him. There are two incidents in which the lawyers confront Jesus about his interpretation of the biblical writings that have much to teach us concerning ethical understanding of gametal donation — one about the unity of marriage and the other about the question "who counts as my neighbor?"

As regards the first issue he says: "Haven't you read that at the beginning the creator made them male and female? . . . and the two will become one flesh? So they are no longer two but one. What God has joined together let man not separate" (Matt. 19:4-7). Note the relevance of the flesh, of our bodily condition for the meaning of the married unity. This instruction is given to help us to understand the sacredness of the married union, its divine origin, its true beginning. It is a unity we are not to disrupt, break down, or separate.

Who is my neighbor? Jesus answers this question with one of the most beautiful stories of the gospel, the Good Samaritan (Luke 10:25-37). He affirms with the story a founding truth of our humanity, of "what and who we are," namely, the preciousness of every human being in the eyes of God, and hence, for each other. We can put this truth in more philosophical terms as "the intrinsic worth and dignity of every human being." The Samaritan's heart "was filled with pity" when he saw his fellow human being "robbed, stripped, beaten, half dead" — wounded in his flesh and soul. The first thing he does to assist him is to care for the wounds of his flesh. He cleans them and bandages them. He then takes him where he can be cared for and pays for all expenses. Our first responsibility to care for our neighbor arises and is immersed in his or her bodily needs. Our neighbor, as Paul Ramsey puts it, is "a person who within the ambience of the flesh claims our care."[2] We are called to assist our neighbor as best we can, whoever he or she may be. The order of the flesh, and the care for the flesh within that order, is essential in human existence. Each one of us is made in God's likeness, which means in Jesus' likeness, as it was "in the beginning."

Based on this biblical understanding, there are three fundamental truths concerning the question "what and who are we?" that give ultimate sense to an ethical appraisal of gametal donation. They may be expressed in three terms: creation, marriage, and the dignity of the person. They can also be expressed in three theses:

(a) There is a natural order, an *order of matter,* an order of material wholeness in the cosmos and in our planet, which bears God's intentional design of which we are part.

(b) Human beings participate in that order of matter simply because they are creatures of flesh. There is an *order of the flesh,* of human embodiment, of human beings as gendered and engendering beings, which, according to God's design, finds fulfillment in the unity of marriage. Our relations to one another, as man and woman in one flesh, is an order of the unity of two persons — as ensouled bodies and embodied souls — a sacred unity that should not be infringed upon, separated, broken down, or destroyed.

(c) There is an *order of neighborliness,* of kindness, of affection, of love for one another as fellow human beings, an order of God's design, enshrined in our embodiment. In this order of affection and care we are expected to live, each "loving our neighbor as we love ourselves." This order demands our attention if we are to survive as well as to honor God.

2. Paul Ramsey, *The Patient as Person: Explorations in Medical Ethics* (New Haven and London: Yale University Press, 1970), p. xiii.

One of the connections between these three dimensions of human existence is their common ground on matter, their material order, distinct but interrelated. It is this material order that technology directly affects. Matter or the natural order is where technology has its greatest impact. For this reason it is important to make the natural order of matter and of our embodiment central to our conversation about forms of technological generation. The deeper significance of this claim will become apparent as the present discussion progresses. Technology impacts humanity, that is, what and who we are as humans, because its purpose is to manipulate the matter on which we live, the flesh in which we live, and the bodies through which we relate to one another, with no overall regard for their natural order — divinely established and maintained.

Recognizing the importance of the three dimensions of human existence discussed above is not limited to the Christian community. The truth about the material order of creation, the order of the environment in our planet, of its living wholeness and beauty (and of our dependence on it) is acknowledged today in secular terms in the new ecological movement. Ecology has emerged as a response to our blindness to and abuse of the natural order through the use of technology. Committed ecologists seek to restore and preserve this marvelous natural order. It is beginning to give us, or help us to regain, some sense of our natural grounds. This concern for ecology can be seen as a sign of hope in our morally chaotic technological culture. Without making us "nature worshipers," it serves us as an antidote to the disregard for the natural order of the world, of its beauty and wholeness. Unfortunately, the ecological movement has not yet touched the inner realms of the human body, the order of the flesh. There is relatively little activist attention being paid to "body-ecology."

The truth about the dignity of the person is expressed and advocated, even if in partial terms, in laws and some current movements defending human rights (for example, Amnesty International). Recently, the president of the United States has stated that a universal goal of our common humanity is "to give an opportunity to every one to develop his or her full potential." Nevertheless, throughout the whole world the unborn stage of human potential, and the potential of each unborn, is not advocated even in principle; rather, it is bracketed out, and largely legally denied. In secular terms, the "full potential" is considered only after birth.

It is hard to identify any secular movement or group that advocates and defends the meaning of marriage as an exclusive union of man and woman, and of the family as founded on marriage. There is no secular movement that advocates the natural order of the flesh, the order of our gendered and engendering embodiment, on the same scale as the ecological movement pursues its cause. God's point of view, in this respect, seems to be present in the world largely through those who are faithful to his vision as handed down through biblical understanding. For most of those who are faithful to God, this is one, *if not the single greatest*, challenge of our time: to live, preserve, and hand down the meaning of our humanity enshrined in a household and a family founded on the biblical understanding of marriage. For where there is no household and no family, there is no future for humanity.[3] The true love between man and woman as one flesh, on which our humanity depends, cannot flourish without them.[4]

Gametal Donation

The following advertisement was posted in a visible spot at a bus stop near the University of Chicago in mid-1998: "Healthy Women Needed. Excellent Compensation." In smaller print was the following: "Healthy women 20-33 needed to serve as anonymous egg donors. Donors will be required to take medication, blood screening, and undergo minor surgical procedure. Substantial compensation will be given. If interested, call ARR (773.327.7315), serious inquiries only. We are interested in all ethnic backgrounds." A call to ARR ("Alternative Reproductive Resources") revealed that the compensation involved was $3,500.

Twenty years earlier a similar advertisement was placed in a London Medical School seeking male medical students to donate their sperm. The compensation then was £7 (about $10 U.S.) per sample. The compensation today in the United Kingdom, as determined by a national regulatory authority, is $40 U.S. plus other expenses the person may have incurred in the donation.[5] In the United States the compensation is about $150 to $200 per sample. A recent conversation about these matters with a doctoral student at the University of Chicago yielded the following remark: "Oh, yes, this is a practice on the campus. I know one of the male students who has paid for his

3. See H. G. Gadamer, "Bodily Experience and the Limits of Objectification," in *The Enigma of Health* (Cambridge: Polity Press, 1996), pp. 70-82. He draws our attention there for the need to recover the paradigmatic importance of the household for the preservation of our humanity.

4. See the chapter by Charles Sell in the present volume.

5. *U.K. Human Fertilisation and Embryology Act,* 1990, HMSO, London. This act established the Human Fertilisation and Embryology Authority (HFEA) as a national regulatory body. Since its inception the Authority has produced an annual report.

graduate studies by regularly donating semen." The studies at the University of Chicago are expensive. So the literal truth of the statement can be questioned, but not its central claim. The student is known to have donated his sperm regularly, a number of times, and the inducement was the financial compensation. This circumstance accords with the findings of a recent European report: "The present survey shows that donors in almost all 30 European countries are initially attracted by the opportunity to earn money."[6]

Conceiving children by someone else's sperm (traditionally referred to as AID, artificial insemination by donor, though now often called intrauterine insemination) has been a widespread practice since the 1950s, when freezing sperm was made possible. As a result, sperm storage and sperm banks were started. The motivations for such procedures have been a combination of eugenic, feminist, and marital demands.[7]

The eugenic intent of seeking "the best," "most perfect," and "most healthy" standard of human beings (to be guaranteed through their bodily frame — the best access science and technology have to the project) is manifested in the sperm bank opened in California in 1980. The project was designed by Robert K. Graham, who wanted to collect the sperm of Nobel Prize–winning scientists. The sperm was to be made available to women of high IQ. It is known that William Schockley, Nobel Prize winner for physics in 1956, contributed to the sperm bank.[8]

In 1983 a sperm bank was opened in Oakland, California, by women dedicated to the feminist cause. All women, regardless of race, genetic status, sexual orientation, or marital status, could avail themselves of the services. Similar initiatives are supported in Europe: "One of the basic human rights is that of a woman to decide when and how to conceive. Under the European Convention, a single woman or even a lesbian couple is entitled to have children, even though these children may not have a legal father."[9]

Donated sperm is used for the purpose of conceiving children through-

---

6. Joseph G. Schenker, "Assisted Reproduction Practice in Europe: Legal and Ethical Aspects," *Human Reproduction Update* 3, no. 2 (1997): 173-84, 177. A complementary report to this one can be found in D. Evans and M. Evans, "Fertility, Infertility and Human Embryo: Ethics, Law and Practice of Human Artificial Procreation," *Human Reproduction Update* 2 (1996): 208-44. There are two large volumes devoted to this same information: D. Evans and Martinus Nijhoff, eds., *Conceiving the Embryo and Creating the Child* (1996), 1: 358 and 2: 366.

7. See Andrew C. Varga, "Eugenics and the Quality of Life," in *The Main Issues in Bioethics* (Paulist Press, 1984), p. 95.

8. *Time*, "Superkids," 10 March 1980, p. 49.

9. European Convention, 1978; quoted in Schenker, "Assisted Reproduction Practice in Europe," p. 175.

out the world — within hospital settings supported by National Health Services, in private clinics, and at home.[10] As the procedure is relatively simple, there are "do-it-yourself" kits for the purpose. The simple form of the procedure consists in introducing seminal fluid into the woman's genital tract, through the vagina, by means of a syringe or catheter.[11]

It is estimated that the number of children born in the United States by gametal donation is 25,000-30,000 annually. In the last four decades, approximately a million children have been born in this way, and two million parents have children by this method. The estimated population in Europe is about 600 million, approximately double the size of the population of the United States, and the number of children born there by gametal donation is greater as well. On top of this, "AID is used extensively throughout the world."[12] The overall number of donors, doctors, parents, and children involved in this mode of generation is huge.

A more recent avenue for gametal donation is in vitro fertilization (IVF). It remained an experimental procedure for two decades, from 1958 until a child was born by this method on July 25, 1978. In this procedure, eggs are isolated and surgically retrieved from the human female body, in order to attain fertilization in the lab with retrieved sperm from a man. In the last twenty years, the IVF method, with all its allied procedures, has encouraged egg donation to become widespread, and with it, the donation of embryos.

There are many countries in Europe in which gametal donation is carried out with no specific legislation to regulate the practice. All the countries adhere to the principle of anonymity of donors, except Sweden. "The Council of Europe recommends that all precautions should be taken to keep secret the identity of all the parties involved, and the identity of the donor must never be revealed even in court."[13] Sweden does not abide by this ruling and has determined that it is the child's right to obtain information about the identity of

---

10. Practices of AID have already been critically assessed in the United Kingdom in the early 1980s by R. Snowden and G. D. Mitchell in *Artificial Family: A Consideration of Artificial Insemination by Donor* (London: Unwin Books, 1981).

11. This procedure was followed by the persons involved in the court case *Baby M,* 1998, New Jersey Lexis, 79; New Jersey Supreme Court No A-39 (February 1988). Case reported by Mary Briody Mahowald, *Women and Children in Health Care: An Unequal Majority* (Oxford University Press, 1996), p. 106. The following is quoted there: "the threesome finally handled the matter themselves. Debbie purchased a diabetic syringe at the drug store and filled it with her husband's sperm. Following the directions in a family medical guide, she successfully injected the sperm into her friend Sue, a twenty-four-year-old virgin, and their child was born nine months later."

12. See Schenker, "Assisted Reproduction Practice in Europe," p. 175.

13. Ibid., p. 177.

the donor. Although the protection of anonymity is still widespread, it has been ethically and legally contested for a long time. The present author challenged it at a hearing of the European Parliament in 1985.[14] Currently, there is not an overall European legislative agreement regarding this matter. In the United States, the degree of regulation varies widely by state.

The People Involved

What vision of ourselves is there, inherent in our technological culture, that promotes the practice of gametal donation as something reasonable, acceptable, or even commendable and altruistic? How do the donors, the clinical staff, prospective parents, and policymakers who advocate, participate in and carry out these practices see and justify themselves as ethical agents? How do they perceive their responsibility to God (if they are believers), to their fellow human beings, and especially to the children conceived by gametal donation, so many of whom are now adults?

There are two dimensions of the practice of gametal donation which are, and have been from its beginning, constitutive aspects of it. I will consider them as two windows through which we can look at the practice and thereby construct an overall ethical evaluation of it. These two aspects are suggested by the previously mentioned "Alternative Reproductive Resources" advertisement's references to "anonymous egg donors" and "substantial compensation." The issues are *anonymity* and *money*.

What does anonymity and the offering and receiving of substantial compensation disclose about the ethical attitudes and self-understanding of those involved in these practices, and of the practices themselves? Our moral attitudes are best manifested in what we do and how we act, not merely in what we say. So what do the actions of those involved in gametal donation, who condone and practice both financial reward and anonymity, tell us about their ethical worldview? How do they understand marriage, family, and our responsibility to one another as human beings, as neighbors? To these issues I now turn. I will consider them as they relate to donors, clinical teams, prospective parents, children, and society.

---

14. See Teresa Iglesias, "European Legislation on Questions Concerning Artificial Insemination: A Report to the European Parliament," in *IVF and Justice* (London: Linacre, 1990), pp. 139-59.

Donors

It is hard to imagine that young healthy women at a university (or anywhere else) would go through the procedure of donating their eggs — having to take medication when they are healthy, and having to undergo surgery while they are healthy — if substantial financial compensation were abolished. Accordingly, the very act of recruiting young women in this manner, and with the health risks involved, is abusive. It puts those in financial need under enormous pressure. Consider the words of a student at Case Western Reserve University who donated eggs: "I would never go through this [drug-induced hyperstimulation and egg retrieval] if I weren't paid $1,000. After all, I'm a poor student."[15] The male student who paid much of his way through graduate school by donating his sperm would most likely have not done so without financial compensation.

The financial side of gametal donation is crucial to its survival. Publicly we are somewhat reluctant to use the terms "buying" and "selling" gametes. Legally, the human body and its parts are not marketable entities; they are not legally regarded as material or chattel to buy and sell (although the project to patent body genes and DNA is defended by a large body of scientists and their allied pharmaceutical companies).[16] Nevertheless, we are forced to recognize the commercialization of gametal donation when someone who requires gametes from a clinic for conception has to pay for them. Part of our body is now marketable material. "'Ova donation,' as we know it, plainly amounts to 'eggs for sale.'"[17]

Receiving $3,500 for one donation of female gametes is described as "substantial." In itself the quantity appears to be so and, therefore, is highly tempting to a student or anyone else who is in need. At the same time, this figure is understandable in the whole context of costs involved in the procedure. A couple requiring one egg for conceiving a child will be asked in the United Kingdom today to pay £1,000 for the egg. In the United States the price of one egg to conceive a child in a clinic typically varies from $1,000 to $2,000.

15. Mahowald, *Women and Children in Health Care*, p. 104. The author states in a footnote: "The student took my course in 'Moral Problems in Medicine' at Case Western Reserve University in 1987. After undergoing drug-induced hyperstimulation of her ovaries, this student underwent laparoscopic removal of her ova" (122). Today, non-surgical removal of ova through the vagina is used as well.

16. Stephen Sherry, "The Incentive of Patients," in *Genetics Ethics: Do the Ends Justify the Genes?* (Grand Rapids: Eerdmans, 1997), pp. 113-23.

17. Mahowald, *Women and Children in Health Care*, p. 104.

The healthy woman who approaches ARR to donate eggs, if she undergoes the required procedures, will have to take medication (a high dose of hormones) for multiple ovulation to occur. About fifteen to thirty ova could be induced to mature and will then be retrieved from her ovaries by minor surgery. If only fifteen eggs are retrieved from the donor, and the clinic uses them all, then $15,000-$30,000 will be ultimately received for the donation. In monetary terms, the difference between what the donor receives from and what she produces for the clinic is also "substantial." Naturally, the costs related to the clinical staff, medication, technology, insurance, and so forth would have to be covered from this differential. These are only rough calculations. But the point is this: financial success has become the driving force.

The donor enters the process of technological generation as a contributor of "reproductive material" (this is a term in current use, even by Christian authors)[18] and as the origin of a "reproductive resource." This is the open admission of those who seek the donation, and of those who give it and receive it. The donor is important only because his or her body is a reproductive resource; it produces the gametes, "cells," from which a child might be conceived. The donor as a person, as a human being, remains faceless, anonymous; he or she is irrelevant. The fact that a donor has a name, a history, and a life story about who he or she is, means relatively little in this mode of generation. The child's origin is taken to be from the gamete — an "anonymous" cell — not really from a parent, as the person who engenders. The genetic connection to the one who possesses the engendering powers, the actual parent, is deemed inconsequential. When the donor gives the gametes, he or she dissociates and is dissociated from them. Once the gametes are given away, they have no more connection with the donor. In this whole process, what matters are the gametes, not the donor as a person; what matters is the relevant "reproductive material" from which a child might develop. The donor as a person and the "reproductive material" are severed. The human body and the human will go separate ways. Human wholeness is broken down.

We can therefore speculate that if the gametes could be produced in the laboratory, the donors would become dispensable, totally irrelevant to the process. Under this presumption, once gametes are made in the lab, people could choose the preferred gametes according to the specifications given in a "Catalogue of Reproductive Resources."

For a gamete donor, parenthood or being a parent is reduced to being "a reproductive source," the originator of "reproductive material." The donor

18. See, for example, John R. Williams, *Christian Perspectives in Bioethics* (Ottawa: Novallis, 1997), p. 75.

parents as persons, who and what they are, are dissociated from the gametes themselves, and hence dissociated from their own bodies; human flesh is reduced to tissue and material. This explains why ultimately it is of no importance what origin the anonymous gametes may have — as long as they are healthy and match familial specifications. Since the end of the procedure is the child to be conceived and born, and this is the desired and ultimate good, the persons who are the source of the gametes can be bracketed, stripped of their familial roles, and reduced to nothing more than to providers and distributors of raw material.

Why is anonymity even at issue here? Why is it still so necessary? Why could not the donors reveal their identity? Why can the prospective parents and future child not know the identity of the donors? Anonymity is required because the responsibilities for the consequences of giving gametes to conceive children are impossible to bear. The principle of anonymity could be translated as "satisfaction of desire without burden of consequences." The satisfaction (to be obtained, by various parties involved, from money, professional status, or the child) must be realized, but its natural consequences negated, abolished, because of their moral weight. Anonymity may be compared to the stage curtain in a play. When it is pulled up, it reveals the finished attractive scene, but when it is down it hides the procedures by which the beautiful scene is made possible. It is our ethical responsibility to remove the curtain to see the complete picture, the whole truth, that is, to see the process and its consequences going on behind the scenes.

Consider some of these consequences:

- Donating gametes makes the donor the genetic parent of perhaps ten conceptions, ten children. He or she thereby becomes related to ten other persons as a genetic parent, and possibly to ten distinct families. (Ten donations are the number of donations permitted by national regulation in United Kingdom since 1990. There was no regulation before. Many countries still have no regulations. But even when some form of regulation exists, there are no means to fully control the number of times a donor donates. The clinics have to trust the donor's word. We can also assume that since the gametes cost money, they are not going to be discarded too quickly.)

- A child conceived by gametal donation will be born to one, and perhaps two parents, who have no gametal link to the child. They could have more children by this method, and perhaps more by using their own gametes. The child could be related to some of the other siblings, but then may be related to none. The child could also be related to one parent or may be related to neither. In the latter case, the child will probably never know much about his or her genetic parents.

- Each child could have other half-brothers and -sisters in different families, permitting the possibility of intermarriages of half-brothers and -sisters. We can also think of the possibility of intermarriage of full brothers and sisters when, coincidentally, gametes from the same donors may be used to generate embryos, which are then donated.

These are stark realities: a donor rejects parental responsibility for all his or her children who are to be conceived and reared by strangers. We could put the matter into a question: Is it humanly possible for a single donor to assume true responsibility, or establish any form of relationship, with numerous children and their parents and families? And if this is not possible, why do people donate? Who would want to be publicly known as the genetic father or mother of numerous children all scattered and unknown? How could a donor publicly advocate any form of parental responsibility and then negate it by becoming a donor? How can our society permit such an extraordinary genetic parental mismatch of attitudes and actions? Only anonymity can make this possible. It hides and it negates. By negating the bond between the genetic parent and the child, the bond is believed not to exist. But genetically, materially, in the flesh, it does exist, and it is known to exist. It is only denied.

However, our denial does not wipe it away. It will eventually surface. It will burst like a bubble. In Britain there is at present a concern about what to tell the children conceived by gametal donation when the year 2008 arrives. They will have reached the age of eighteen and will have the legal right to consult records about their parental origins. (This right was established by an act of Parliament implemented on August 1, 1990; nevertheless under the act, legally accessible records are still minimal.) To solve these deep moral predicaments, counseling is advocated: "We must hope, whatever information is to be divulged, that it will be offered in a 'sensitive and considerate manner'. . . . The importance of counseling, enshrined in the U.K. Act, will be paramount."[19]

If there were no financial compensation, would there be donors of gametes? Perhaps some, but very few. If there were no anonymity, would there be donors of gametes? Perhaps some, but very few. This is the experience in Sweden. Another telling example from Britain can be added, from a source revealing information about egg donation at the Middlesex Hospital in Lon-

---

19. H. Abdalla, E. Shenfield, and E. Latarche, "Statutory Information for the Children Born from Oocyte Donation in the U.K.: What Will They Be Told in 2008?" *Human Reproduction* 13, no. 4 (1998): 1106-9.

don.[20] From 1991 to 1997 a total of 585 women donated eggs. Of these, 389 (66.7 percent) were anonymous; 196 (33.3 percent) were known donors (who donated directly to relatives or friends). But of all women, only 6 percent answered an optional question on the "Confidential Donor Information Form" which asked the women to "give some description of yourself as a person." It is interesting to ponder why so many women were unwilling to respond to a question regarding their very person, regarding who they were.

To whom is anonymity beneficial? we may ask. At first sight to the donors, then to the physicians, and finally to the prospective parents. The donors would not want any liability for the child's welfare, nor any claims to inheritance rights. Physicians would not want responsibilities for any medical mishap. The parents would want to protect "family privacy." Yet it is recognized by all that anonymity is not beneficial to children. They are deliberately deprived of the knowledge and meaning of their genetic origins. They are deliberately deprived of important knowledge about genetic information that may have medical significance. They are deliberately denied any link to their familial human origins.

Ethically, I want to argue that anonymity is not beneficial to anyone. Those who defend anonymity suggest that it is meant to serve the longings of infertile couples to have a child of their own. For this reason they see it as justified. "Anonymity serving desire" can be taken as its outer ethical layer, that is, the layer of sympathy — wanting to see the couple happy by having their desire for a child satisfied. Yet, there is a deeper ethical layer to anonymity beyond serving desire and its illusions; it also serves the consciousness that something is not right, that something is amiss, and for this reason it must remain covered, hidden to oneself and to others.[21]

Deep down, anonymity manifests a great violation of our humanity; it is itself inhuman. It does not build on true human goods and relationships. It betrays truth and the basic trust we owe to each other, to our children, and to ourselves. After the stage is set and the curtain is pulled, we may see happy parents interacting with their healthy children (of course, this is not always so). But when we remove the curtain of anonymity and are able to watch the stage being set, we see the inhumanity of the process. The process:

- induces the participants — prospective parents and donor — to live in pretense (as if they were/were not parents); it encourages self-deception and the deception of others.

20. Ibid.
21. Leon R. Kass in conversation suggested the ideas in this paragraph.

- satisfies partial, not integral, desires, that is, it does not do justice to the good and true human desires of everyone concerned, including the children.
- permits donors to abandon the responsibility of human bonding with their genetic offspring.
- makes donors and their agents appear to be responsible and honorable by remaining anonymous.
- burdens female donors with financial pressure and serious health risks related to the retrieval of eggs (involving multiple ovulation, use of anesthesia, and surgical procedures) plus possible damage to the ovaries.
- breaks basic marital, parental, and familial bonds and commitments.
- relies on the natural grounds of personal human embodiment, where the marriage and the family are based, and yet negates and destroys those grounds in order to construct a "new family" constituted by roles determined by rational will.[22]
- denies children dignity equal to that of other children and adults by denying them their roots, familial heritage, and bonds, all of which are needed to help us establish our identity.
- covers up genetic relationships, which could result in marriages between half-brothers and half-sisters.
- may result in necessary medical genetic history being unobtainable.
- promotes the commercialization of the human body.

The Clinical Teams

In 1985 the matter of anonymity in gametal donation was discussed at a public hearing in one of the commissions of the European Parliament. A French physician, whose practice involved administering AID, openly stated: "If anonymity is abolished we will have no donors, and I will burn all my records." This doctor, by his own decision and consent, will have patients ten years later with no accurate medical genetic familial history. The accessible records kept by doctors are still "minimal," not detailed, regarding the physical, personal, and medical history of the donor. One of the U.S. president's ethics commissions concerned with this matter has stated: "When AID results in genetic

22. The implications of creating an "artificial family" are treated by M. Luisa Di Pietro, "Artificial Fertilisation and the Family," *Proceedings of the International Conference of Women's Health Issues* 1997, Rome and Washington (forthcoming). I thank the author for allowing me to consult her paper and profit from it in unpublished form.

105

disease, the source of the sample cannot be determined; semen from the donor may be used again and result in another child with the disease. Indeed the commission heard testimony about just such a case."[23]

Scientists and doctors have made technological generation a possibility and a reality. A mode of generation that involves coming into existence asexually, that is, by technological means, truly is a "reproductive revolution." (The term "re-production" is nearer — certainly in cloning — to what technology makes of human generation; but it is not a truly human term; human beings are neither "productions" nor "reproductions" of their parents.) The major revolutionary novelty in human generation is not that human beings can have sex without children (for this is a natural human event in many situations without the Pill). The revolutionary technological novelty is that human beings can have children without sex. Technology has put human generation totally outside the power of our bodily human nature and its intrinsic purposes. With gametal donation, which is an inherently asexual form of technological generation, the act of engendering has become more of a technologically constructed act than a truly human act that respects human beings as personal and bodily engendering beings. Leon Kass describes this profound reality most adequately: "Human procreation . . . is not simply an activity of our rational wills. It is a more complete activity precisely because it engages us bodily, erotically, and spiritually, as well as rationally. There is wisdom in the mystery of nature that has joined the pleasure of sex, the inarticulate longing for union, the communication of the loving embrace, and the deep-seated and only partially articulate desire for children in the very activity by which we continue the chain of human existence and participate in the renewal of human possibility. Whether or not we know it, the severing of procreation from sex, love, and intimacy is inherently dehumanizing, no matter how good the product."[24]

Gametal donation, like all asexual reproduction, fosters objectification and promotes the transformation of the order of material nature in the name of progress. As H. G. Gadamer so wisely denounces, it demands, as a result, "a violent estrangement from ourselves," since as unique bodily and personal individuals we cannot be objectified.[25] The modern objectifying scientific outlook induces us to see our bodily nature as what this technological culture

23. *President's Commission for the Study of Ethical Problems in Medicine and Biomedical and Behavioral Research, Screening, and Counseling for Genetic Conditions* (Washington, D.C.: U.S. Government Printing Office, 1983), p. 11. Reported in Varga, "Eugenics and the Quality of Life."

24. See Kass, "Wisdom of Repugnance," p. 22.

25. See Gadamer, "Bodily Experience," p. 70.

calls it: "reproductive material." As already indicated, this objectification is manifested in the practices of the clinical team in their obtaining and handling of donor gametes and their adherence to anonymity.

Gametal donation, being a mode of technological generation, is really not a way of assisting nature; it is rather a form of activity whose major project is changing and using nature, our bodily nature, in order to get what we want from it. Modern scientific knowledge is instrumental knowledge, technical knowledge, particularly in the realm of human (and animal) generation. It seeks to understand nature (material nature and its powers) in order to intervene in it. But the intervention is not primarily for the purpose of helping nature to maintain its wholeness or to become more whole in its order and powers. Scientific-technological intervention aims at making nature do what by its own powers and finalities it cannot do. In this sense, technology is not holistic. It is not integrative and respectful of integrity. Many of its advances come at the expense of the disintegration and destruction of matter and its order. That is why the ecological movement has emerged as a response to recover the constant loss of wholeness and integrity in the material environment, and in our understanding of nature as an ordered whole.

By providing gametal donation, clinical teams have become body technicians, the "technical experts" of the human body, dedicated to altering the natural order of procreation, responding to the wants and desires of their clients. "Confronted with the growing demand of donated oocytes and the diminished number of donors, clinicians all over the world have suggested to use ovarian tissues from live donors, cadavers, or from aborted fetuses as potential sources of female gametes for donation."[26] Note the word "clinicians." Why have clinicians thought of such a suggestion? (So far the suggestion has not been taken up.) It must be said that due to some advances in technology, a large portion of modern medicine has lost its goal of assisting and healing nature. The clinicians seem to have lost a sense of any natural and human boundaries. Nature, including humanity, is not understood by these clinicians to have an internal order, a wholeness and purpose which is intrinsic to it. Bodily, we are envisaged as "useful matter," the *res extensa* of Descartes, to be used for the rational willful "I," the *res cogitans* Descartes and his followers tell us human beings are. In the Cartesian perspective (a rationalistic, scientific, and dualistic perspective), inert matter is at the service (mercy) of the desires of our rational will. The human being is thus split, divided, cut in two, denaturalized and rationalized rather than healed and made whole.

In technological generation it is doctor-scientists who "harvest" (to use

26. See Schenker, "Assisted Reproduction Practice in Europe," p. 179.

their language) the gametes. The doctor-scientist evaluates, discards, mixes, and makes the product with the material at his or her disposal. In this process, "quality control" demands elimination of diseases by eliminating the defective matter (which includes embryos and fetuses). This eugenic intent is built into the IVF programs through pre-implantation diagnosis, genetic diagnosis, and/or selective reduction by abortion. The IVF program in many centers works in conjunction with gametal donation.

For many doctor-scientists, technological generation and money go hand in hand. They have made and continue to make the relationship between donors and clients possible. Along with the conceptions and births of children by these procedures, practitioners manage a business that is financially rewarding not only for donors, but for themselves as well.

Prospective Parents

Prospective parents have to recognize that in entering a program of technological generation by gametal donation they become an active part of the whole process, humanly, financially, and ethically. All its participants make the practice as a whole, in all its dimensions, possible.

Prospective parents may entertain the possibility of gametal donation for various reasons. The basic one with which we all sympathize is "to have a child of their own" — a drive well examined by Gilbert Meilaender in the present volume. But they agree to have a child who is genetically unrelated to one or both of them. The natural desire to have a child is usually located by the married couple, and those who support the procedure, within the overall goods of life and family. The couple seeks:

- the good of life giving in the spousal context;
- the good of caring and bringing up children;
- living in family, with others, responsible for their well-being, witnessing with joy their growth and development;
- a "normal family" with two parents, father and mother, and their children;
- the fulfillment of their own humanity and that of others in family life.

Yet the fulfillment of these natural and very human desires is really "partial" when the child is not the fruit of the parental union.

The major ethical issue in gametal donation for married couples, nevertheless, is the meaning and reality of being married. How does the couple understand being married or getting married? Would it make sense at the time of entering marriage, and making marriage a lifelong commitment, to say to the prospective spouse: "I take you as my wife (or husband) and will remain so if we can have a child, or if you can give me a child." This would mean that the decision for union is conditional. There would be no marriage with this conditional intention. One partner would be instrumentalizing the other.

The reality of marriage has its roots in what and who we are as bodily personal beings, as gendered and engendering beings — as God created us. To bring forth new life requires the union of man and woman as bodily and personal beings. Such a union of two is exclusive. It is a personal mutual commitment. By this commitment each one becomes fully dedicated and entrusted to the other, as one flesh, in all dimensions of their being: bodily, emotionally, morally, spiritually. The unifying bond of mutual love and self-giving in all these dimensions is constitutive to marriage.

It is through marrying that man and woman make each other a spouse. The woman makes the man her husband. The man makes the woman his wife. This is the great truth: man and woman are made spouses through a commitment to one another. Equally, both spouses are made parents through each other. The husband makes his wife a mother when she conceives and bears a child, and she makes her husband a father when she conceives and bears a child. In their union, the spouses make each other fertile, life giving. True fertility is a matter of two, not of each spouse separately. Husband and wife are fertile or infertile together. For only man and woman together, not separately, are able to conceive a child.

The child conceived is the natural gift of the spouses who are engendering beings committed to each other. Natural spousal conception occurs by means of what the spouses are and who they are, in their human wholeness and totality, not only in will and plan. The will does not conceive; it is the body that conceives. Autonomous decisions and choice by themselves do not cause conception. Intimate bodily union does.

To see a spouse as infertile, and hence as the failure to be "corrected" in the engendering powers of a couple, is to reduce him or her to nothing more than the source of "reproductive material." Even if one partner would consent to gametal donation, such consent violates the married union. No longer are the spouses becoming parents by virtue of their total union and commitment to one another. God's design for human procreation is broken down. Consider the difficulty of having to tell children of their origins. They will have to learn and understand their procreation process from donated gametes. They will have to deal with what it means to be wanted by a social parent and not by a genetic parent whom the child may never come to know. The wisdom written in our nature is truly wisdom — not mere "neutral processes" that we can happily disregard. Married couples should be fertile, become fertile, or remain infertile together. The intentional introduction of a third person as genetic parent (as opposed to a redemptive, after-the-fact adoption) violates the sacredness of the union of two persons in marriage; it forces a redefinition of family that is not in harmony with our nature and not in harmony with the revealed will and design of God for us.

Children

What is the impact on children of this mode of generation? Much has already been said; but this can be added. We are storied beings, storied selves. Each one is a life, with a beginning, a middle, and an end. Each one has his or her story, which constitutes his or her meaningful identity: what to tell to others — and oneself — about oneself. This story has many layers and facets. "I am so and so," we often tell each other. What is the content of "so and so"? It is a brief part of our story. Our gender identity is stated by our very presence. Only we may know other parts of our inner stories, the deeper layers. Yet communication, as the key to personal intimacy, is sharing and constructing our stories together, of becoming *who we are* together. This is the only way we can be and become who we are. Much of our story is written in our bodies, the parents, the grandparents, the brothers, the sisters, we have. In the United States we are even asked for our ethnicity in semi-official documents. A child conceived by gametal donation is only and deliberately provided with a *partial story*. The story begins somewhere, but no one knows where, because everyone hides it. When it is not fully hidden, even the full story cannot be told. And this is deliberately planned, constructed, clinically assisted, legally approved, and consented to by everyone participating in the process — everyone, that is except the child.

What is the child to make of having been conceived from "the reproductive resources" of a clinic? How is the child to interpret that he or she may have ten or more half-brothers and half-sisters scattered throughout the city or the country? Will not this be relevant for them when they may want to marry? Will they not want to investigate through DNA their genetic origins? How are they to respond to what they learn about the donors, clinical team, and parents? If they are not told, why have their lives been planned by others to be lived out half-storied or in deception? How are they supposed to interpret the reality of their conception as the best possible way for them to have come into the world? How are they to think of a society that places the wants and desires of "parents" above those of their own? Who are they to trust fully?

From an ethical point of view, we should not have gametal donation at all, and hence it should not be clinically assisted. Minimally, from a legal point of view to ameliorate the present situation, we must not deprive a child of access to his or her genetic origins. It is an injustice to do so. At present, even if the parents want to inform the child of his or her form of conception and historical heritage, or if the child wants to learn more about his or her origins, complete records do not exist. They are deliberately not kept. Is it not a great paradox that in a culture of choice, this obvious choice is deliberately denied to some children? The principle "in order to avoid trouble keep people ignorant of the truth of their predicament" does not work long-term. It has never worked. Injustice, whatever form it takes, always catches up with us, for it bears within it the crying claim for justice. From a Christian point of view, how are we to interpret "walk in the truth" and "the truth will make you free"? What is it to be faithful to God?

Society

A decent society cannot base its social, professional, and familial relationships (involving donors, clinical teams, parents, children, brothers and sisters, larger extended family, and friends) on a refusal to disclose truth and demand responsibility, and a willingness to betray trust with deception. What simple minimal recommendation could be suggested for a social improvement of the situation? (1) We should remove financial compensation: donors should truly "donate" (i.e., gratuitously). (2) We should remove anonymity. (3) We should obtain and keep accurate records and have full confidential information available to those directly involved. (4) There should be legal instruments whereby children conceived by gametal donation are effectively protected and treated as equal in rights and dignity as other children and adults. "Then we will have no donors!" as the French physician said. But then, that is exactly what we should have.

*   *   *

Why is gametal donation so widespread — and yet so hidden from public view — that even children conceived by this method do not know it themselves? The answer that has been given throughout our discussion is that the core of gametal donation is covered by anonymity, and the practice would either end or radically change if anonymity were totally removed — together with its alliance to money. This is so, because anonymity manifests (by hiding) that *in this mode of generation something is humanly and morally radically amiss.*

Gametal donation is dehumanizing and outside of God's design for us. It breaks the sacredness of the unity and wholeness of marriage, of parenthood, and of bodily integrity. It regards our engendering powers as mere "reproductive resources." It treats children as instruments of adults' desires. Overall, it violates the meaning and reality of what we are as engendering beings, and of who we are as personal beings, called to live by responsible truthful commitments to ourselves, to our spouses, to our children, to others, and to God.

CHAPTER 7

Surrogate Motherhood

Scott B. Rae, Ph.D.

In 1998, one of the most unusual cases of surrogate motherhood came before the courts, a case that is older than the child who is at its center.[1] Five separate people contributed to the conception, birth, and upbringing of one child. John and Luann Buzzanca "created" a child through in vitro fertilization using donor eggs and sperm. They hired a surrogate mother to carry the child to term. After the child's birth (on April 26, 1995), the Buzzancas intended to raise the child, whom they named Jaycee. Two weeks after the Buzzancas signed a contract with the surrogate, they separated. John filed for divorce one month prior to the child's birth. At the divorce hearing, the court ruled that John had to pay child support, which he later appealed. The surrogate filed for custody of the child — a petition she later dropped. In May 1997, a superior court ruled that Jaycee has no parents and that John was not obligated to pay child support, thus leaving the little girl in legal limbo. That decision was reversed in March 1998, when the court of appeals ruled that the Buzzancas are the parents and that John is responsible for child support. This case illustrates the legal and moral complexities of surrogate motherhood that are still being sorted out.

Surrogacy arrangements are morally troublesome for a Christian who holds to a creation model for procreation, since such a model suggests skepticism toward any procreative arrangement that utilizes a third party contributor — of eggs, sperm, or wombs.[2] However, there are other issues that are

---

1. *Buzzanca v. Buzzanca* (1998 Cal. App. Lexis 180), 10 March 1988. For commentary on the case see Scott B. Rae, "Surrogate Mothers Must Have Rights, Too," *Orange County Register*, 13 March 1998, p. B5.

2. For more on this, see Scott B. Rae, *Brave New Families: Biblical Ethics and Reproductive Technologies* (Grand Rapids: Baker Book House, 1996), pp. 23-36.

unique to surrogacy, such as the charge of baby-selling and the thorny debate over the definition of motherhood in surrogacy arrangements. These issues must be considered if the Christian is to address policymakers and bring a Christian ethic for surrogacy into a secular society. These matters will be the focus of this chapter.

A wide variety of surrogacy arrangements are possible today. One form is called *genetic surrogacy,* in which the surrogate contributes both the egg and the womb to the couple who contracts her services. The surrogate is genetically related to the child she is carrying. In some cases, the infertile woman can produce eggs but cannot carry a pregnancy to term. Then what is called *gestational surrogacy* is needed. This is a more complicated and expensive approach, since in vitro fertilization is also required. The surrogate has no genetic tie to the child she is carrying. When the surrogate is paid a fee for the entire process, it is called *commercial surrogacy.* Most surrogates are paid for what they do, somewhere between $10,000 and $15,000 in addition to all medical expenses. Sometimes even lost wages due to the pregnancy are reimbursed. But once in a while, a family member or close friend offers to carry a child for an infertile woman simply out of a desire to give the "gift of life." When this happens, it is called *altruistic surrogacy.*

Commercial Surrogacy Is Baby-Selling

Commercial surrogacy is morally and legally problematic because it involves the sale of children. As of this writing, twenty-five states currently have laws that prohibit remuneration for the adoption of a child.[3] These laws were en-

3. States which have such laws include Alabama (Ala. Code 26-10-8, 1977); Arizona (Ariz. Rev. Stat. Ann. 8-126 (c), 1974); California (Cal. Penal Code 273 (a) West 1970); Colorado (Colo. Rev. Stat. 19-4-115, 1973); Delaware (Del. Code tit. 13, 928, 1981); Florida (Fla. Stat. Ann. 63.212(1)(b), West Supp. 1983); Georgia (Ga. Code Ann. 74-418, Supp. 1984); Idaho (Idaho Code 18-1511, 1979); Illinois (Ill. Rev. Stat. ch. 40, 1526, 1701, 1702, 1981); Indiana (Ind. Code Ann. 35-46-1-9, West Supp. 1984-85); Iowa (Iowa Code Ann. 600.9, 1981); Kentucky (Ky. Rev. Stat. 199.590(2), Supp. 1986); Maryland (Md. Ann. Code 5-327, 1984); Massachusetts (Mass. Ann. Laws ch. 210, 11A, Michie Law Coop. 1981); Michigan (Mich. Comp. Laws Ann. 710.54, West Supp. 1983-84); Nevada (Nev. Rev. Stat. 127.290, 1983); New Jersey (N.J. Stat. Ann. 9:3-54, West Supp. 1984-85); New York (N.Y. Soc. Serv. Law 374(6), McKinney 1983); North Carolina (N.C. Gen. Stat. 48-37, 1984); Ohio (Ohio Rev. Code Ann. 3017.10(A), Baldwin 1983); South Dakota (S.D. Codified Laws Ann. 25-6-4.2, 1984); Tennessee (Tenn. Code Ann. 36-1-135, 1984); Utah (Utah Code Ann. 76-7-203, 1978); and Wisconsin (Wisc. Stat. Ann. 946.716, West 1982). Some states, such as Arizona, California, and Florida, exempt stepparents from such laws. Ap-

acted to prevent economically and emotionally vulnerable birth mothers from being coerced into giving up children for adoption. The abuses and excesses of black-market adoptions were, and still are, the target of these laws. Most of the states that have passed surrogacy laws have applied existing adoption law to surrogacy and prohibited commercial surrogacy.

Most proponents of surrogacy are very sensitive to the charge of baby-selling and insist that the fee paid to the surrogate is for her gestational services rendered. They maintain that the fee is simply another expense for the contracting couple, parallel to the medical and legal expenses involved. Most surrogacy contracts today are structured to relate the fee to the specific gestational services rendered by the surrogate, and those who frame the contracts are careful not to make any mention of the fee paying for the transfer of parental rights, the most critical component in the arrangement. For example, Karen Marie Sly terms surrogacy prenatal baby-sitting, not baby-selling, and argues that the surrogate has the right to rent her womb for a fee.[4]

The argument that the surrogacy fee is for services rendered and not for the sale of a child fails to take into account both the nature of the surrogacy contract and the intended end of a surrogacy arrangement. Most surrogacy contracts are structured around the product, not the process, or the service of surrogacy. For example, in the Baby M case, the Stern-Whitehead contract specified that only in the event that Whitehead delivered a healthy baby to the Sterns would she be paid the entire $10,000 fee. If she miscarried prior to the fifth month of pregnancy, she would receive no fee, though all medical expenses would be paid. If she miscarried after the fifth month, or if the child was stillborn, she would only receive $1,000 of the fee. The contract was oriented to delivery of the end product, not the service rendered in the process.[5] Normally, the majority of the fee (usually half), if not all of it (as was the case with the Stern-Whitehead case), is withheld until parental rights are actually

---

proximately half of the states impose criminal sanctions ranging from simple misdemeanor to felony. Some will not grant a final disposition of the adoption proceedings until a list of expenses incurred in connection with the adoption has been submitted to the court (Arizona, Delaware, Iowa, Michigan, New Jersey, and Ohio).

4. Karen Marie Sly, "Baby-Sitting Consideration: The Surrogate Mother's Right to 'Rent Her Womb' for a Fee," *Gonzaga Law Review* 18 (1982-83): 549-51.

5. William F. May suggests that Judge Sorkow's distinction in the lower court Baby M decision between the process and product of surrogacy "flies in the face of the facts." He states: "Let no one doubt that the contract points toward the delivery of a product. And when the mother delivers that child, she must relinquish her rights as its genetic mother, her presumptive rights of custody over an already extant child. It is baby-selling." "Surrogate Motherhood and the Marketplace," *Second Opinion* 9 (1987): 135.

waived and the custody of the child is turned over to the contracting couple. Thus, it is difficult to see how the fee can be for gestational services only when the service itself is not the final intent of the contract. Payment is made upon the surrogate fulfilling all the necessary responsibilities to insure transfer of parental rights. Alexander M. Capron and Margaret J. Radin suggest that the claim that the fee is for gestational services alone is merely a disguise that serves to hide the true intent of the contract. They state:

> The claim that the payment to the surrogate is merely for "gestational services" is just a pretense, since payment is made "upon surrender of custody" of the child and for "carrying out . . . , obligations" under the agreement. These include taking all steps necessary to establish the biological father's paternity and to transfer all parental rights to the biological father and his mate.[6]

Accordingly, contracting couples sue for specific performance of the contract when the surrogate decides to keep the child. If the contract were for gestational services only, on what basis would the couple request the state to enforce the terms of the contract, when the surrogate "breaches" it by keeping the child? Under the gestational services scheme, specific performance could only refer to the surrogate's failure to maintain the pregnancy, either by abortion on her demand, or by engaging in personal behavior, such as drinking, smoking, or drug use, that brings harm to the fetus. Thus specific performance could only relate to the procreative services for which the contracting couple is presumably paying. Any damages which would be awarded in lieu of specific performance (courts typically prefer to award damages than to enforce specific performance in service contracts) would be groundless, since the surrogate would have not actually breached the contract.

Other proponents of commercial surrogacy grant that it does constitute baby-selling, but argue that surrogacy is so different from the black-market adoptions at which the adoption laws are aimed that they should not apply to surrogacy. For example, in surrogacy, the natural father of the child is also the intended social father of the child. Thus, how can the natural father buy back what is already his? However, this argument erroneously assumes that the child for which the natural father has contracted, and for which he is paying, is *all* his. It clearly is not. The surrogacy contract is parallel to a quit claim deed in a piece of property, in which he buys out the surrogate's interest in the child, making it a commercial transaction for the rights to the child, or baby-selling.

6. A. M. Capron and M. J. Radin, "Choosing Family Law over Contract Law as a Paradigm for Surrogate Motherhood," *Law Medicine and Health Care* 16, no. 1-2 (1988): 37.

A second claimed difference between surrogacy and black-market adoptions is that the latter are only concerned with money, not the child's best interests. This claim assumes that all surrogacies are about the best interests of the child, which is an overstatement at best. The only screening that occurs for the contracting couple is financial, and their fitness to be parents is rarely considered.[7]

A third suggested difference is that surrogacy results from a planned and often desperately desired pregnancy, not an unwanted pregnancy. Thus the surrogate cannot be coerced into giving up her child out of desperation, as is the case in many unwanted pregnancies that end up as adoptions. However, this fact does not mean that surrogacy is free from coercion, since in some cases the surrogate does develop a desire to keep the child after she enters the contract. She may well have a wanted child and unwanted contract.

The strongest argument against commercial surrogacy is based on the violation of human dignity that occurs when any human being is an object of barter. Since commercial surrogacy clearly involves the sale of children, it is prohibited on principle. Human beings, and particularly the most vulnerable human beings (children), are simply not for sale, even to benevolent buyers. People have inherent worth that places them above the vicissitudes of the market. In the West, this affirmation has ultimately been grounded theologically in the dignity and value of the individual, who is the image of God and thus cannot be for sale. It is also grounded constitutionally in the Thirteenth Amendment, which prohibits slavery principally on the grounds that human beings, irrespective of race or color, are not items that can be traded for financial consideration. It is further grounded in the almost universally accepted Kantian maxim that people are ends, not merely means, and should be treated as such. There are also consequential reasons for prohibiting the sale of human beings, especially children. For example, such a prohibition prevents exploitation of the mother and the contracting couple, safeguards the best interests of the child from the profit motive, prevents lack of counseling and screening of the contracting couple, and insures that predictable human suffering is prevented.

7. As George Annas comments:

Surrogacy brokers need a dose of realism. They should not be permitted to hide behind the grief of infertile couples. They are not in business to help them; they are in business to make money. The primary screening brokers do is monetary; does the couple have the $25,000 fee?

See George J. Annas, "Death Without Dignity for Commercial Surrogacy: The Case of Baby M," *Hastings Center Report* 18 (April-May 1988): 22.

117

Beyond all of these considerations, though, the most basic reason that human beings should not be bought and sold is that such practices inherently violate personal integrity and dignity, by regarding people as something they intrinsically are not. The New Jersey Supreme Court, for example, has recognized that baby-selling is problematic outside of its potential for exploitation. Appealing to the inherent worth of persons, they state, "There are, in a civilized society, some things that money cannot buy. . . . There are . . . values that society deems more important than granting to wealth whatever it can buy, be it labor, love or life."[8] The sale of children, which inevitably results from a surrogacy "transaction," is inherently problematic, irrespective of the goods that emerge, in the same way that slavery is inherently morally troubling — because human beings are not objects that are for sale.

Definition of Motherhood in Surrogacy Cases

Some of the most vexing questions in surrogacy revolve around the issue of parental rights. With the ability to separate the genetic, gestational, and social components of motherhood successfully, new questions are being raised for which society has few legal, moral, or social precedents. The presumption has always been that biology determines parenthood. As a result, society has never been forced to examine the social and moral aspects of motherhood. There is general agreement that in cases in which the surrogate provides both the egg and the womb, she is both the natural and legal mother, and thus possesses full maternal rights. The heart of the present debate concerns gestational surrogacy, in which two women make biological contributions. One donates the egg and the other "rents the womb." Genetics, gestation, and preconception intent to parent all have been proposed as the determinant of motherhood in these arrangements.[9]

A Christian perspective on this debate would suggest the following. When a couple has their own sperm and egg joined in a laboratory and an embryo results, they are the parents of that embryo (embryonic child). Accordingly, they remain the parents unless they agree to give up that status to someone else. (The use of donor eggs and sperm will not be addressed at this point but is discussed in the present volume's chapter by Teresa Iglesias.) It has been easy for many to conclude, then, that if the parents decide to use a

---

8. In *re Baby M,* 537 A. 2d, 1249 (1988).

9. For further discussion of these positions, see Scott B. Rae, *The Ethics of Commercial Surrogate Motherhood* (Westport, Conn.: Praeger Publishing, 1994), pp. 77-124.

surrogate for the pregnancy and intend to remain the parents, the woman involved is simply providing a service to the parents and does not have any genuine maternal relationship with the developing child. Such a construal far underestimates the significance of the gestational mother (here, the surrogate) and her relationship with the child,[10] and wrongly implies that the only relevant intent is the preconception intent of the genetic parents. We need to examine the issues of gestation and intent more carefully here before formulating a conclusion about parental status when surrogacy is involved.

The gestational mother is anything but a "human incubator" and makes a substantial contribution not only to the physical development of the child, but to his or her emotional and psychological development as well. The gestational mother has not simply donated the eggs; she has built up what Ruth Macklin calls "sweat equity" in the child she is carrying.[11] The nine months invested in the child and the labor literally involved in giving birth are substantial contributions. She clearly has made the greater investment in the child in terms of effort and time expended, and thus her claim to motherhood is understandable. At the end of the process, the woman who gives birth to the child will have contributed much more of herself than the egg donor in order to bring about the child's birth. For a woman who knows what pregnancy and childbirth involve, the contribution of the egg donor might even seem trivial compared to the rigors and demands of the "around the clock" nature of pregnancy and birth.

George Annas underscores the importance of the gestational mother because of the biological investment being made.[12] Though this investment is difficult to define and more difficult to quantify, it does reflect the substantial difference in involvement between egg donation and pregnancy/birth.[13] It is

---

10. The American College of Obstetricians and Gynecologists goes so far as to state, "The genetic link between the commissioning parent(s) and the resulting infant, while important, is less weighty than the link between surrogate mother and fetus, or infant that is created through gestation and birth." *Statement on Surrogate Motherhood* (Washington, D.C.: ACOG, 1990), p. 2. Also cited in Larry Gostin, ed., *Surrogate Motherhood: Politics and Privacy* (Bloomington: Indiana University Press, 1990), pp. 300-303.

11. Ruth Macklin, "Artificial Means of Reproduction and Our Understanding of the Family," *Hastings Center Report* 21 (January-February 1991): 9.

12. George J. Annas, "Redefining Parenthood and Protecting Embryos: Why We Need New Laws," *Hastings Center Report* 14 (October 1984): 51. See also George J. Annas and Sherman Elias, "Noncoital Reproduction," *Journal of the American Medical Association* 255 (3 January 1986): 67; and Annas and Elias, "In Vitro Fertilization and Embryo Transfer: Medicolegal Aspects of a New Technique to Create a Family," *Family Law Quarterly* 17 (1983): 216-17.

13. Laurence D. Houlgate, "Whose Child? In re Baby M and the Biological Preference Principle," *Logos (USA)* 9 (1988): 167.

simply not accurate to suggest that the "carrier" of the child has no impact on the person into whom the child develops. Though the physical traits and many of the predispositions of the child have their source in the genes, there is a growing body of evidence that points to the gestational environment as a substantial contributor to the child's personality. In other words, the significance of the nine-month investment of the gestational woman, including the pain, risk, and sacrifice involved in carrying and giving birth to a child, is considerable.[14]

Similar in importance to the investment and contribution of the gestational mother is the bonding that occurs between her and the child she is carrying. Though significant bonding can take place between any two individuals, the combination of biological investment and the resultant bonding is a potent combination.[15] In most pregnancies, this bond is a central part of the pregnant woman's self-concept, and though children do not normally entirely define a woman's life, they are surely integral to who she is as a person. In most instances when a pregnancy is lost, the loss of this bond causes a great deal of grief. Such is even the case in adoption, in which the birth mother realizes that giving up the child is in both the child's and the mother's best interests. One reason that many states have a period in which a birth mother can regain custody of her child prior to the adoption's finalization is that they recognize the strength of this bond. Similarly, most states do not hold an adoptive mother to a pre-birth consent to adoption, since she cannot know the strength of the bond she will feel with her child prior to birth, and thus cannot give genuinely informed consent. Though one should be careful about affirming a "biology is destiny" concept of self for women, pregnancy, childbirth, and motherhood can be highly influential in a woman's sense of self. The sense of bonding is what makes pregnancy essentially a relationship, not only physiologically, but also emotionally. A unique relationship is developed with the fetus in the nine months of pregnancy and birth.[16]

---

14. Katharine Bartlett, "Re-expressing Parenthood," *Yale Law Journal* 98 (1988): 329-30.

15. See "Rumplestiltskin Revisited: The Inalienable Rights of Surrogate Mothers," *Harvard Law Review* 99 (June 1986): 1952.

16. Barbara Katz Rothman, "Surrogacy Contracts: A Misconception," *Daily Journal Report* 88-6 (1 April 1988): 20. Rothman summarizes the argument for gestation in this way: "Any pregnant woman is the mother of the child she bears. Her gestational relationship establishes her motherhood. We will not accept the idea that we can look at a woman, heavy with child, and say the child is not hers. The fetus is part of the woman's body, regardless of the source of the egg and sperm. Biological motherhood is not a service, not a

It is true that the contracting couple anticipate a very strong bond with the child when they gain custody. In many surrogacy cases, they participate in the medical exam visits and viewing of ultrasound images of the child. Unless the relationship between the surrogate and contracting couple is adversarial from the start, the couple is not usually isolated from the experience of the surrogate. However, participating in a pregnancy as an observer is very different from actually carrying the child. Even though the contracting couple may be anticipating a strong bond with the child when they take over custody, that bond is anticipated and not currently experienced in the same way as the gestational mother experiences a bond during pregnancy.

The surrogate mother, then, has a far stronger maternal relationship with the child in utero than does the genetic mother. The significance of this consideration is often underappreciated, while the importance of the genetic mother's preconception intent is often overestimated. There is no reason why the surrogate's developing intent to parent should be regarded as less important than the preconception intent of the genetic parents. After all, the surrogate has experienced a relationship with the developing child, and her strong attachment to the child she is carrying is the result of the bonding that is characteristic of pregnancy. At this point in the pregnancy, the surrogate combines both the biological involvement and the intent to parent the child in a way that is not comparable to the genetic parents. Simply because the intent to become a parent develops during the pregnancy is no reason why it should be given less consideration, particularly since the developing intent is grounded in the bonding and relationship with the child whom the surrogate is carrying.

Once we more accurately appreciate the strong relationship between the surrogate mother and the child she is carrying, and more clearly recognize the inadequacy of placing too great an emphasis on preconception intent, we are in a better position to address the question of parental status. Carrying a child through a pregnancy is so substantial an undertaking and forges such a strong relationship between surrogate and child that no woman ought to be asked to do so without being acknowledged as the legal mother of

---

commodity, but a relationship." It should be noted that pro-choice feminists like Rothman cannot have it both ways. One cannot hold that pregnancy is essentially a relationship and support liberal abortion laws at the same time. If a woman wants to end a relationship with her unborn child, she has the option of adoption, not abortion. A right to end a relationship does not give a woman the right to end the life of the one with whom she wants to end the relationship.

the child. Just as the genetic mother actually turns over her child to another woman to be the mother during the pregnancy, so the law should explicitly affirm that such a transfer of maternal status takes place in a gestational surrogacy arrangement. Since entering into this type of surrogacy arrangement would remain voluntary, it would simply be understood to necessarily entail voluntarily transferring legal maternal status from the genetic mother to the surrogate. All of the stringent requirements normally a part of adoption proceedings — informed consent, a waiting period, and so forth — would apply. The intention of such an arrangement, of course, would be that the surrogate mother would give up the child at birth for adoption by the genetic mother. But this latter adoption would also have to follow all of the normal adoption guidelines, including the stipulation that it be truly voluntary on the part of the surrogate mother.

It should immediately be apparent that such an approach to surrogacy, while corresponding best to the mother-child relationships that actually do exist, is very risky for the genetic parents. There is a chance they may end up losing custody of their genetic child. To be sure, lawyers who arrange surrogacy contracts frequently testify that in many successful surrogacy arrangements no bond is established between surrogate and child, even when there is a genetic connection. The child is turned over to the contracting parents and there is no dispute concerning custody. But even if mother-child bonding is not an entirely universal experience, and if, parallel to adoption, surrogates do often successfully relinquish custody, that is one thing. It is quite another to insist that the contract be fulfilled and the child turned over to the genetic parents when the gestational mother has indeed established a significant bond with the child. In those cases in which a breach of the agreement occurs, denying legal motherhood to the woman who carries the child runs contrary to the powerful combination of biological and relational connections. If the surrogate desires to relinquish the child to the couple who initiated the surrogacy arrangement, then she is free to do so. But all must recognize that the child is hers, as the legal mother, to give or to keep.

It is hard to imagine that couples in general — and Christian couples in particular — would consider this approach to surrogate motherhood a reproductive option worth pursuing. Not only would they risk losing custody of their genetic children, but they would be bringing about a situation in which they were encouraging a pregnant mother — indeed, the mother of "their own" child — to do precisely the opposite of what God created her to do. She was created to pour herself, physically and emotionally, into her developing child. Yet she would be asked to refrain from at least the emotional dimension of the investment she would normally make in the child.[17] If all dimensions of our being are important parts of an integrated whole, such an arrangement would undoubtedly prove destructive for the child and for the woman herself, to a degree that, at present, God only knows.

*   *   *

Surrogate motherhood arrangements are likely to be with us for the foreseeable future. However, Christian couples should be skeptical of any procreative arrangements that involve a third-party contributor. Public policy can prove helpful to Christians and others alike in two ways. It can make commercial surrogacy illegal, and it can clarify that motherhood status is legally as well as functionally transferred during pregnancy to surrogate mothers, who need only transfer that legal status back if they choose to do so.

17. For example, a book that advertises itself as a leading "consumer's guide" to new reproductive technologies underscores the importance of this. It states, "The surrogate must begin to erect an emotional barrier . . . , so that she experiences the child not as hers but as the child of the couple." In fact, there is a growing body of literature produced by the surrogacy industry that gives instruction to potential surrogates on how to do this during the course of their pregnancy. See Lori Andrews, *New Conceptions* (New York: Ballantine, 1985), p. 221. Daniel Callahan suggests that should surrogacy become widespread, "We will be forced to cultivate the services of women with the hardly desirable trait of being willing to gestate and then give up their own children. . . . There would still be the need to find women with the capacity to dissociate and distance themselves from their own child. This is not a psychological trait we should want to foster, even in the name of altruism." See Daniel Callahan, "Surrogate Motherhood: A Bad Idea," *New York Times,* 20 January 1987, p. B21.

CHAPTER 8

Human Cloning

John F. Kilner, Ph.D.

We live in a brave new world in which reproductive technologies are ravaging as well as replenishing families. Increasingly common are variations of the situation in which "baby's mother is also grandma — and sister."[1] Sometimes extreme measures are necessary in order to have the kind of child we want.

This new eugenics is simply the latest version of the age-old quest to make human beings — in fact, humanity as a whole — the way we want them to be: perfect. It includes our efforts to be rid of unwanted human beings through abortion and euthanasia. It more recently is focusing on our growing ability to understand and manipulate our genetic code, which directs the formation of many aspects of who we are, for better and for worse.

We aspire to complete control over the code, though at this point relatively little is possible. This backdrop can help us understand the great fascination with human cloning today. It promises to give us a substantial measure of power over the genetic makeup of our offspring. We cannot control their code exactly, but the first major step in that direction is hugely appealing: You can have a child whose genetic code is exactly like your own. And you didn't turn out so badly, did you?

Admittedly, in our most honest moments we would improve a few things about ourselves. So the larger agenda here remains complete genetic control. But human cloning represents one concrete step in that direction, and the forces pushing us from behind to take that step are tremendous. These forces are energized, as we will see, by the very ways we look at life and

1. Bette-Jane Crigger, ed., *Cases in Bioethics,* 2nd ed. (New York: St. Martin's Press, 1993).

justify our actions. But before examining such forces, we need a clearer view of human cloning itself.

The Rising Prospect of Human Cloning

It was no longer ago than 1997 when the president of the United States first challenged the nation — and charged his National Bioethics Advisory Commission[2] — to give careful thought to how the United States should proceed regarding human cloning. Attention to this issue was spurred by the reported cloning of a large mammal — a sheep — in a new way. The method involved not merely splitting an early-stage embryo to produce identical twins. Rather, it entailed producing a nearly exact genetic replica of an already existing adult.

The technique is called "nuclear transfer" or "nuclear transplantation" because it involves transferring the nucleus (and thus most of the genetic material) from a cell of an existing being to an egg cell in order to replace the egg cell's nucleus. Stimulated to divide by the application of electrical energy, this egg is guided by its new genetic material to develop into a being that is genetically almost identical to the being from which the nucleus was taken. This process was reportedly carried out in a sheep — to produce the sheep clone named Dolly[3] — but attention quickly shifted to the prospects for cloning human beings (by which I will mean here and throughout, cloning by nuclear transfer).

Quickly people began to see opportunities for profit and notoriety. By 1998, for example, scientist Richard Seed had announced intentions to set up a Human Clone Clinic — first in Chicago, then in ten to twenty locations nationally, then in five to six locations internationally.[4] While the U.S. federal government was pondering how to respond to such initiatives, some of the states began passing legislation to outlaw human cloning research, and nineteen European nations acted quickly to sign a ban on human cloning itself.[5] However, the European ban only blocks the actual implantation, nurture, and

---

2. See National Bioethics Advisory Commission, *Cloning Human Beings: Report and Recommendations of the National Bioethics Advisory Commission,* June 1997.

3. Ian Wilmut et al., "Viable Offspring Derived from Fetal and Adult Mammalian Cells, *Nature* 385 (1997): 810-13.

4. Peter Kendall, "Image of Human Cloning Proponent: Odd and Mercurial," *Chicago Tribune,* 11 January 1998, p. 6.

5. "Europe Moves to Ban Human Cloning," *Bulletin of Medical Ethics,* January 1998, pp. 3-5.

birth of human clones, and not also cloning research on human embryos that are never implanted. Such research has been slowed in the United States since the president and then Congress withheld federal government funds from research that subjects embryos to risk for non-therapeutic purposes.[6] Moreover, a United Nations declaration co-sponsored by eighty-six countries in late 1998 signaled a broad worldwide opposition to research that would lead to human cloning.[7]

Yet there are signs of this protection for embryos weakening in the face of the huge benefits promised by stem cell research. Stem cells can have the capacity to develop into badly needed body parts such as tissues and organs. The easiest way to produce stem cells is to divide an early stage embryo into its component cells — thereby destroying the embryonic human being. The National Institutes of Health has decided that as long as private sources destroy the embryos and produce the stem cells, the federal government will fund research on those cells.[8] These developments underscore that there are a number of technological developments that are closely interrelated and yet have somewhat different ethical considerations involved. While embryo and stem cell research are very important issues, they are distinct ethically from the question of reproducing human beings through cloning. Reproduction by cloning is the specific focus of this chapter.

While no scientifically verifiable birth of a human clone has yet been reported, the technology and scientific understanding are already in place to make such an event plausible at any time now. There is an urgent need to think through the relevant ethical issues. To begin with, is it acceptable to refer to human beings produced by cloning technology as "clones"? It would seem so, as long as there does not become a stigma attached to that term that is not attached to more cumbersome expressions like "a person who is the result of cloning" or "someone created through the use of somatic cell nuclear transfer." We call someone from Italy an Italian — no disrespect intended. So it can be that a person "from cloning" is a clone. We must be ready to abandon this term, however, if it becomes a "label" that no longer meets certain ethical criteria.[9]

---

6. President Clinton issued his directive to the National Institutes of Health on 2 December 1994, and congressional action (PL104-91/PL104-208) took effect with the fiscal year 1996 budget.

7. United Nations Commission on Human Rights, *Universal Declaration on the Human Genome and Human Rights* (approved on 19 November 1998).

8. Rick Weiss, "NIH to Fund Controversial Research on Human Stem Cells," *Washington Post*, 20 January 1999, p. A2. See ethical critique at *www.stemcellresearch.org*.

9. Labels "must be precisely and relevantly defined. They must be accurately applied. And they must lead to treatment that serves the welfare of those that are labeled." See

Why Clone Human Beings?

In order to address the ethics of human cloning itself, we need to understand why people would want to do it in the first place. People often respond to the prospect of human cloning in two ways. They are squeamish about the idea — a squeamishness Leon Kass has argued we should take very seriously.[10] They also find something alluring about the idea. Such fascination is captured in a variety of films, including *The Boys from Brazil* (portraying the attempt to clone Adolf Hitler), *Bladerunner* (questioning whether a clone would be more like a person or a machine), and *Multiplicity* (presenting a man's attempt to have enough time for his family, job, and other pursuits by producing several live adult replicas of himself). Popular discussions center on the wonderful prospects of creating multiple Mother Teresas, Michael Jordans, or other notable figures.

The greatest problem with creative media-driven discussions like this is that they often reflect a misunderstanding of the science and people involved. The film *Multiplicity* presents human replicas, not clones in the form that we are discussing them here. When an adult is cloned (e.g., the adult sheep from which Dolly was cloned), an embryo is created — not another adult. Although the embryo's cells contain the same genetic code as the cells of the adult being cloned, the embryo must go through many years of development in an environment that is significantly different from that in which the adult developed. Because both our environment and our genetics substantially influence who we are, the embryo will not become the same person as the adult. In fact, because we also have a spiritual capacity to evaluate and alter either or both our environment and our genetics, human clones are bound to be quite different from the adults who provide their genetic code.

If this popular fascination with hero-duplication is not well founded, are there any more thoughtful ethical justifications for human cloning? Many have been put forward, and they cluster into three types: utility justifications, autonomy justifications, and destiny justifications. The first two types reflect ways of looking at the world that are highly influential in the United States and elsewhere today, so we must examine them carefully. Although they could be analyzed in explicitly Christian terms, such language will be avoided here in order to illustrate ways that popular approaches can be critiqued on their own terms. The third, while also influential, helpfully opens the door to

---

Ralph B. Potter, "Labeling the Mentally Retarded: The Just Application of Therapy," in *Ethics in Medicine,* ed. Stanley J. Reiser et al. (Cambridge, Mass.: M.I.T. Press, 1977), pp. 626-31.

10. Leon R. Kass, "The Wisdom of Repugnance: Why We Should Ban the Cloning of Humans," *Valparaiso University Law Review* 32 (spring 1998): 679-705.

more explicitly theological reflection. I will begin by explaining the first two justifications. In the following sections I will then assess the first two justifications and carefully examine the third.

Utility

Utility justifications defend a practice based on its usefulness, or benefit. As long as it will produce a net increase in human well-being, it is warranted. People are well acquainted with the notion of assessing costs and benefits, and it is common to hear the argument that something will produce so much benefit that efforts to block it must surely be misguided. Even Christian analysis has sometimes sanctioned this utilitarian way of thinking by equating it with fulfilling the biblical command for us to love our neighbors.[11]

Utility justifications are common in discussions of human cloning. Typical examples include:

1. By having clones, people can, in some measure, have more of themselves in the world and thereby make a bigger impact.
2. Parents can replace a dying child with a genetically identical new one.
3. Parents can produce a clone of a sick child to provide bone marrow or other lifesaving bodily elements that can be provided with relatively modest risk to the clone.
4. Parents, both of whom have a lethal recessive gene, can produce a child by cloning rather than risk the one-in-four chance that their child will face an early death.
5. Clones could be produced to provide organs for transplant — admittedly, transplants that could jeopardize or even end a clone's life.
6. Other clones could be produced with unusually high or low mental capacities that would suit them well to do socially needed tasks — for example, challenging problem solving or menial labor.

Autonomy

The second type of justification appeals to the idea of autonomy — an increasingly popular appeal in this postmodern age, in which people's personal

---

11. For example, see the classic example of Joseph Fletcher, *Situation Ethics* (Philadelphia: Westminster Press, 1966).

experiences and values play a most important role in determining what is "right" and "true" for them. According to this justification, we ought to respect people's autonomy as a matter of principle. People's beliefs and values are too diverse to adopt any particular set of them as normative for everyone. Society should do everything possible to enhance the ability of individuals and groups to pursue what they deem most important.

Again, there are many forms that autonomy justifications can take. However, three stand out as particularly influential in discussions of human cloning:

1. *Personal freedom.* There is a strong commitment in many countries, the United States in particular, to respecting people's freedom. This commitment is rooted in a variety of religious and secular traditions. Respect for people entails allowing them to make important life decisions that flow from their own personal values, beliefs, and goals, rather than coercing them to live by a burdensome array of social requirements.
2. *Reproductive choice.* Reproductive decisions are especially private and personal matters. They have huge implications for one's future responsibilities and well being. Social intrusion in this realm is particularly odious.
3. *Scientific inquiry.* A high value has long been placed on protecting the freedom of scientific inquiry. More knowledge and better understanding enhance our capacity to make good decisions and accomplish great things in the world.

Utility and autonomy are important ethical justifications. However, they do not provide a sufficient ethical basis for human cloning. We will examine them here carefully in turn.

Understanding Utility

While the concern for utility is admirable, there are many serious problems with this type of justification. Most significantly, it is *unworkable* and it is *dangerous*. It is unworkable because knowing how much utility cloning or any other practice has, with a reasonable level of precision, is simply impossible. We cannot know all of the ways that a practice will affect all people in the world infinitely into the future. For example, it is impossible to accurately quantify the satisfaction of every parent in future centuries who will choose cloning rather than traditional sexual reproduction in order to spare their children from newly discovered genetic problems that are now unknown. In fact, as sheep cloner Ian Wilmut was widely quoted as observing, shortly after announcing his cloning of Dolly, "Most of the things cloning will be used for have yet to be imagined." The difficulty of comparing the significance of every foreseeable consequence on the same scale of value — including comparing each person's subjective experiences with everyone else's — only adds to the unworkability.

What happens in real life is that decision makers intuitively compare only those consequences they are most aware of and concerned about. Such an approach is an open invitation to bias and discrimination, intended and unintended. Even more dangerous is the absence of limits to what can be justified. There are no built-in protections for weak individuals or minority groups, including clones. People can be subjected to anything — the worst possible oppression or even death — if it is beneficial to the majority. Situations such as Nazi Germany and American slavery can be justified using this way of thinking.

When utility is our basis for justifying what is allowed in society, people are "used," fundamentally, as mere means to achieve the ends of society or of particular people. It may be appropriate to use plants and animals in this way, within limits. Accordingly, most people do not find it objectionable to clone animals and plants to achieve products that will fulfill a purpose — better milk, better grain, and so forth. However, it is demeaning to "use" people in this way.

This demeaning is what bothers us about the prospect of producing a large group of human clones with low intelligence so that society can have a source of cheap menial labor. It is also what is problematic about producing clones to provide "spare parts" such as vital transplantable organs for other people. Both actions fail to respect the equal and great dignity of all people by making some, in effect, the slaves of others. Even cloning a child who dies to remove the parents' grief forces the clone to have a certain genetic makeup in order to be the parents' child, thereby permanently subjecting the clone to the parents' will. The irony of this last situation, though, is that the clone will not become the same child as was lost — both the child and the clone being the product of far more than their genetics. The clone will be demeaned by not being fully respected and accepted as a unique person, and the parents will fail to regain their lost child in the process.

To summarize: The utility justification is a substantially inadequate basis for defending a practice like cloning. In other words, showing that a good benefit — even a great benefit — will result is not a sufficient argument to justify an action. Although it is easy to forget this basic point when enticed by the promise of a wonderful benefit, we intuitively know it is true. We recognize that we could, for example, cut up *one* person, take her or his various organs for transplant, and save *many* lives as a result. But we do not go around doing that. We realize that if the action we take to achieve the benefit is itself horrendous, beneficial results are not enough to justify it.

As significant a critique as this is of a utility justification for human cloning, there is more to say. For even if it were an adequate type of justification — which it is not — it is far from clear that it would justify human cloning. To justify human cloning on the basis of utility, *all* the consequences of allowing this practice have to be considered — not only the benefits generated by the exceptional situations commonly cited in its defense. What are some of the consequences we need to be concerned about? There is only space here to note two of the many that weigh heavily against human cloning.

First, as suggested earlier, to allow cloning is to open the door to a much more frightening enterprise: genetically engineering people without their consent — not for their own benefit, but for the benefit of particular people or society at large. Cloning entails producing a person with a certain genetic code because of the attractiveness or usefulness of a person with that code. In this sense, cloning is just the tip of a much larger genetic iceberg. We are developing the genetic understanding and capability to shape the human genetic code in many ways. If we allow cloning, we legitimize in principle the entire enterprise of designing children to suit parental or social purposes. As one researcher at the U.S. Council on Foreign Relations has commented, "Dolly is best understood as a drop in a towering wave (of genetic research) that is about to crash over us." The personal and social destructiveness of large-scale eugenic efforts (including but by no means limited to Nazi Germany's) has been substantial, but at least it has been restricted to date by our limited genetic understanding and technology.[12] Today the stakes are much higher.

The second of the many additional considerations that must be included in any honest utilitarian calculus involves the allocation of limited resources. To spend resources on the development and practice of human cloning is to not spend them on other endeavors that would be more beneficial to society. For many years now there have been extensive discussions about the expense of health care and the large number of people (tens of millions), even

12. See Arthur J. Dyck, "Eugenics in Historical and Ethical Perspective," in *Genetic Ethics: Do the Ends Justify the Genes?* ed. John F. Kilner et al. (Grand Rapids: Eerdmans, 1997), pp. 25-39.

in the United States, that do not have health insurance.[13] It has also long been established that such lack of insurance means that a significant number of people are going without necessary health care and are suffering or dying as a result.[14] Another way of observing similar pressing needs in health care is to survey the specific areas that could most benefit from additional funds.[15] In most of these areas, inadequate funding yields serious health consequences because there is no alternative way to produce the basic health result at issue.

Not only are the benefits of human cloning less significant than those that could be achieved by expending the same funds on other health care initiatives, but there are alternative ways of bringing children into the world that can yield at least one major benefit of cloning — children themselves. If there were enough resources available to fund every technology needed or wanted by anyone, the situation would be different. But researching and practicing human cloning will result in serious suffering and even loss of life because other pressing health care needs cannot be met.

An open door to unethical genetic engineering technologies and a misallocation of limited resources, then, are among the numerous consequences of human cloning that would likely more than outweigh the benefits the practice would achieve. As previously argued, we would do better to avoid attempting to justify human cloning simply based on its consequences. But if we are tempted to do so, we must be honest and include *all* the consequences and not be swayed by exceptional cases that seem so appealing because of the special benefits they would achieve.

Assessing Autonomy

Many people today are less persuaded by utility justifications than they are by appeals to autonomy. While the concern for freedom and responsibility for

13. See discussions in John F. Kilner et al., eds., *The Changing Face of Health Care: A Christian Appraisal of Managed Care, Resource Allocation, and Patient-Caregiver Relationships* (Grand Rapids: Eerdmans, 1998).

14. Office of Technology Assistance, Congress of the United States, *Does Health Insurance Make a Difference?* (Washington, D.C.: U.S. Government Printing Office, 1992).

15. Numerous reports available from the World Health Organization, and UNICEF in particular, document current unmet needs. Projections of U.S. health care expenditures suggest that significant needs in the United States and other countries will persist well into the future. See Office of the Actuary, U.S. Health Care Financing Administration, "The Next Ten Years of Health Spending: What Does the Future Hold?" *Health Affairs* (September-October 1998).

one's own life in this way of thinking is admirable, autonomy justifications are as deeply flawed as utility justifications. More specifically, they are *selfish* and they are *dangerous*.

The very term by which this type of justification is named underscores its selfishness. The word "autonomy" comes from two Greek words, *auto* (meaning "self") and *nomos* (meaning "law"). In the context of ethics, appeals to autonomy literally signify that the self is its own ethical law — that it generates its own standards of right and wrong. There is no encouragement in this way of looking at the world to consider the well-being of others, for that is irrelevant as long as it does not matter to me. Although in theory I should respect the autonomy of others as I live out my own autonomy, in practice an autonomous mindset predisposes me to be unconcerned about how my actions will affect others.

As long as the people making autonomous choices happen to have good moral character that predisposes them to be concerned about the well-being of everyone else, there will not be serious problems. In the United States to date, the substantial influence of Christianity — with its mandate to love others sacrificially — has prompted people to use their autonomous choices to further the interests of others alongside of their own. As Christian influences in public life — from public policy to public education — continue to be eradicated in the name of "separation of church and state," the self-centeredness of an autonomy outlook will become increasingly evident. Consciously or unconsciously, selfish and other base motives arise within us continually, and without countervailing influences, there is nothing in an autonomy outlook to ensure that the well-being of others will be protected.

When autonomy rules, then, scientists, family members, and others are predisposed to act on the basis of their own autonomous perspectives, and the risk to others is real. Herein lies the danger of autonomy-based thinking — a danger that is similar to that attending a utility-oriented outlook. Protecting people's choices is fine as long as all people are in a comparable position to make those choices. But if some people are in a very weak position economically or socially or physically, they may not be able to avail themselves of the same opportunities, even if under more equitable circumstances they would surely want to do so. In an autonomy-based approach, there is no commitment to justice, caring, or any other ethical standards that would safeguard those least able to stand up for themselves.

An autonomy justification is simply an insufficient basis for justifying a practice like human cloning. In other words, showing that a freedom would otherwise be curtailed is not a sufficient argument to justify an action. We have learned this lesson the hard way, by allowing scientific inquiry to proceed unfettered. The Nuremberg Code resulted from research atrocities that were allowed to occur because it was not recognized that there are other ethical considerations that can be more important than scientific and personal freedom (autonomy).[16]

While the autonomy justification itself is flawed, there is more to say about it as a basis for defending human cloning. For even if it were an adequate type of ethical justification — which it is not — it is far from clear that it would actually justify the practice. An honest, complete autonomy-based evaluation of human cloning would have to consider the autonomy of all persons involved, including the people produced through cloning, and not just the autonomy of researchers and people desiring to have clones. Of the many considerations that would need to be taken into account if the autonomy of the clones were taken seriously, space will only permit the examination of two here.

First, human cloning involves a grave risk to the clone's life. There is no plausible way to undertake human cloning at this point without a major loss of human life. In the process of cloning the sheep Dolly, 276 failed attempts occurred — including the death of several so-called "defective" clones. An alternative process used to clone monkeys added the necessary destruction of embryonic life to these other risks. It involved transferring the genetic material from each of the cells in an eight-celled embryo to other egg cells in order to attempt to produce eight so-called clones (or, more properly, identical siblings). Were these experimental technologies to be applied to human beings, the evidence and procedures themselves show that many human embryos, fetuses, and infants would be lost, whatever the process. The loss would be compounded by the fact that it is unlikely human cloning research would be limited to a single location. Rather, similar mistakes and loss of human life would be occurring almost simultaneously at various private and public research sites.

Normally, experimentation on human beings is allowed only with their explicit consent. (Needless to say, it is impossible to obtain a clone's consent to be brought into existence through cloning.) An exception is sometimes granted in the case of a child — including one still in the womb — who has a verifiable medical problem which experimental treatment may be able to

16. Arthur J. Dyck, "Lessons from Nuremberg," in *Ethics in Medicine,* ed. Jay Hollman and John Kilner (Carol Stream, Ill.: Bridge Publications, 1999). See also the classic discussion in Leo Alexander, "Medical Science under Dictatorship," *New England Journal of Medicine* 241 (July 14, 1949): 40-46; cf. Arthur L. Caplan, ed., *When Medicine Went Mad: Bioethics and the Holocaust* (Totowa, N.J.: Humana Press, 1992).

cure or help. However, human cloning is not covered by this exception for two reasons. First, there is no existing human being with a medical problem in the situation in which a human cloning experiment would be attempted. Second, even if that were not an obstacle, there is typically no significant therapeutic benefit to the clone in the many scenarios for which cloning has been proposed. For the experiment to be ethical, not only would there need to be therapeutic benefit to the clone, but that benefit would have to be so huge as to outweigh the substantial likelihood of death or deformity as occurred in the Dolly experiment. To proceed with human cloning at this time, then, would involve a massive assault on the autonomy of all clones produced, whether they lived or died.

There is also a second way that human cloning would conflict with the autonomy of the people most intimately involved in the practice, that is, the clones themselves. Human cloning would radically weaken the family structure and relationships of the clone and therefore be fundamentally at odds with their most basic interests. Consider the confusion that arises over even the most basic relationships involved. Are the children who result from cloning really the siblings or the children of their "parents" — really the children or the grandchildren of their "grandparents"? Genetics suggests one answer and age the other. Regardless of any future legal resolutions of such matters, child clones (not to mention others inside and outside the family) will almost certainly experience confusion. Such confusion will impair their psychological and social well being — in fact, their very sense of identity. A host of legal entanglements, including inheritance issues, will also result.

This situation is problematic enough where a clearly identified family is involved. But during the experimental phase in particular, identifying the parents of clones produced in a laboratory may be even more troublesome. Is the donor of the genetic material automatically the parent? What about the donor of the egg into which the genetic material is inserted? If the genetic material and egg are simply donated anonymously for experimental purposes, does the scientist who manipulates them and produces a child from them become the parent? Who will provide the necessary love and care for the damaged embryo, fetus, or child that results when mistakes are made and it is so much easier just to discard them?

As the U.S. National Bioethics Advisory Commission's report has observed, human cloning "invokes images of manufacturing children according to specification. The lack of acceptance this implies for children who fail to develop according to expectations, and the dominance it introduces into the parent-child relationship, is viewed by many as fundamentally at odds with the acceptance, unconditional love, and openness characteristic of good parenting."[17] "It just doesn't make sense" — to quote Ian Wilmut, who objected strenuously to the notion of cloning humans after he succeeded in producing the sheep clone Dolly.[18] He was joined by U.S. President Clinton, who quickly banned the use of federal funds for human cloning research, and by the World Health Organization, who summarily labeled human cloning "ethically unacceptable."[19] Their reaction resonates with many, who typically might want to *have* a clone, but would not want to *be* one. What is the difference? It is the intuitive recognition that while the option of cloning may expand the autonomy of the person producing the clone, it undermines the autonomy of the clone.

So the autonomy justification, like the utility justification, is much more problematic than it might at first appear to be. We would do better not even to attempt to justify human cloning by appealing to this type of justification because of its inherent shortcomings. But if we are to invoke it, we must be honest and pay special attention to the autonomy of the person most intimately involved in the cloning — the clone. Particular appeals to "freedom" or "choice" may seem persuasive. But if only the autonomy of people other than clones is in view — or only one limited aspect of a clone's autonomy — then such appeals must be rejected.

The Destiny Justification

As noted near the outset of the chapter, there is a third type of proposed justification for human cloning which moves us more explicitly into the realm of theological reflection: the destiny justification. While other theological arguments against cloning have been advanced in the literature to date,[20] many of them are somehow related to the matter of destiny. According to this justification, it is part of our God-given destiny to exercise complete control over our reproductive process. In fact, Richard Seed, in one of his first in-depth interviews after announcing his intentions to clone human beings commer-

17. National Bioethics Advisory Commission, p. 69.

18. He later expanded on his concerns about human cloning in his article "Cloning for Medicine," *Scientific American* 279 (December 1998): 58-63.

19. "WHO Adopts Resolution Against Cloning Humans," Reuters News Service, 16 May 1997.

20. See, for example, the 1998 essays in the journal *Ethics and Medicine* — including those by C. Ben Mitchell (vol. 14:1) and John Grabowski (vol. 14:3). See also the collection of essays in the spring 1998 issue of the *Valparaiso University Law Review* (vol. 32:2), featuring articles by such people as Gilbert Meilaender and Daniel Heimbach.

cially, made this very argument.[21] No less a theologian, President Clinton offered the opposite view when he issued the ban on human cloning. Rather than seeing cloning as human destiny, he rejected it as "playing God."[22] Whether or not we think it wise to take our theological cues from either of these individuals, what are we to make of the proposed destiny justification itself? Is human cloning in line with God's purposes for us?

To begin with, there are indeed problems with playing God the way that proponents of human cloning would have us do. For example, God can take utility and autonomy considerations into account in ways that people cannot. God knows the future, including every consequence of every consequence of all our actions — people do not. God loves all persons equally, without bias, and is committed and able to understand and protect the freedom of everyone — people are not. Moreover, there are other ways that the pursuit of utility and autonomy are troubling from a theological perspective.

The utility of human cloning, first of all, is that we can gain some benefit by producing clones. But using other people without their consent for our ends is a violation of their status as beings created in the image of God. People have a God-given dignity that prevents us from using them as mere means to achieve our purposes. Knowing that people are created in the image of God (Gen. 1:26-27), biblical writers in both the Old and New Testaments periodically invoke this truth to argue that human beings should not be demeaned in various ways (e.g., Gen. 9:6; James 3:9). Since plants and animals are never said to be created in God's image, it is not surprising that they can be treated in ways (including killing) that would never be acceptable if people were in view (cf. Gen. 9:3 with 9:6).

An autonomy-based justification of human cloning is no more acceptable than a utility-based justification from a theological perspective. Some Christian writers, such as Allen Verhey, have helpfully observed that autonomy, understood in a particular way, is a legitimate biblical notion. As he explains, under the sovereignty of God, acknowledging the autonomy of the person can help ensure respect for and proper treatment of people made in God's image.[23] There is a risk here, however, because the popular ethics of autonomy has no place for God in it. It is autonomy *over* God, not autonomy *under* God. The challenge is to affirm the critical importance of respect for human beings — and for their freedom and responsibility to make decisions

21. On the ABC program *Nightline*, 7 January 1998.
22. This language was explicitly affirmed in his 1998 State of the Union address.
23. Allen D. Verhey, "Playing God," in *Genetic Ethics: Do the Ends Justify the Genes?* pp. 60-74.

that profoundly affect their lives — but to recognize that such freedom requires God. More specifically, such freedom requires the framework in which autonomy is under God, not over God — a framework in which respecting freedom is not just wishful or convenient thinking that gives way as soon as individuals or society as a whole have more to gain by disregarding it. It must be rooted in something that unavoidably and unchangeably *is*. In other words, it must be rooted in God — in the creation of human beings in the image of God.

God is the creator, and we worship God as such. Of course, people are creative as well, being the images of God that they are. So what is the difference between God's creation of human beings, as portrayed in the book of Genesis, and human procreation as happens daily all over the world (also mandated by God in Genesis)? Creation is *ex nihilo* — out of nothing. That means, in the first sense, that God did not just rearrange already existing materials. God actually brought into being a material universe where nothing even existed before. However, God's creation *ex nihilo* suggests something more. It suggests that there was no agenda outside of God that God was following — nothing outside of God that directed what were acceptable options. When it came to the human portion of creation, God created us to be the way God deemed best.

It is no accident that we call what we do when we have babies "procreation." "Pro" means "for" or "forth." To be sure, we do bring babies "forth." But the deeper meaning here is "for." We bring new human beings into the world "for" someone or something. To be specific, we continue the line of human beings for God — in accordance with God's mandate to humanity at the beginning to "be fruitful and multiply" (Gen. 1:28). We also create for the people whom we help bring into being. We help give them life, and they are the ones most affected by our actions. What is particularly significant about this "procreation" — this "creation for" — is that by its very nature it is subject to an outside agenda — to God's agenda primarily, and secondarily to the needs of the child being created.

In this light, the human cloning mindset is hugely problematic. With unmitigated pride it claims the right to create rather than procreate. It looks neither to God for the way that he has intended human beings to be procreated and raised by fathers and mothers who are the secondary — that is, genetic — source of their life; nor does it look primarily to the needs of the one being procreated. As we have seen, it looks primarily to the cloner's own preferences or to whatever value system one chooses to prioritize (perhaps the "good of society," etc.). In other words, those operating out of the human cloning mindset see themselves as Creator rather than procreator. This is the kind of aspiring to be God for which God has consistently chastised people, and for which God has ultimately wreaked havoc on many a society and civilization.

Leon Kass has observed that we have traditionally used the word "procreation" for having children because we have viewed the world — and human life in particular — as created by God. We have understood our creative involvement in terms of and in relation to God's creation.[24] Today we increasingly orient more to the material world than to God. We are more impressed with the gross national product than with the original creation. So we more commonly talk in terms of re*production* rather than pro*creation*. In the process, we associate people more closely with things — with products — than with the God of creation. No wonder our respect for human life is deteriorating. We become more like that with which we associate. If we continue on this path — if our destiny is to clone ourselves — then our destiny is also, ultimately, to lose all respect for ourselves, to our peril.

Claims about utility, autonomy, or destiny, then, are woefully inadequate to justify human cloning. In fact, a careful look at any of these types of justification shows that they provide compelling reasons instead to reject human cloning. To stand up and say so may become more and more difficult in our "brave new world." As the culture increasingly promotes production and self-assertion, it will take courage to insist in the new context of cloning that there is something more important. But such a brave new word — echoing the Word of old — is one that we must be bold to speak.

24. Leon Kass, *Toward a More Natural Science* (New York: Free Press, 1985), p. 48.

PART III

DIFFICULT CASES

CHAPTER 9

The Case of the Abortive Surrogate

Introduction

"The Case of the Abortive Surrogate," written and choreographed by C. Christopher Hook, was first enacted as a moot court trial at a conference organized by The Center for Bioethics and Human Dignity in Bannock-burn, Illinois.[1] Among other issues, this case explores the great array of ethical and legal issues that arise when using frozen embryos and a gestational carrier (surrogate mother) to assist reproduction. In the commentaries that follow, Dr. Hook orients us to the issues of the case, C. Ben Mitchell expands on the surrogacy questions involved, R. Martin Palmer and Joyce Shelton examine the proposed abortion, and Samuel Casey considers the legal status of embryos.

This case is based on several actual (though disguised) patient situations. The enactment featured, among others, the authors of the case commentaries in this section of the book. It was presided over by the Honorable W. Dale Young, judge for the Circuit Court of Blount County, Tennessee, who made the landmark ruling in *Davis v. Davis* that the human embryo is a person.

The audience of the moot court served as the jury for the case. At the

---

1. Audio and video versions are available from The Center for Bioethics and Human Dignity (see front of book for contact information).

conclusion of the testimony and arguments, Judge Young asked five questions of the jury:

1. Under the law of East Dakota, there can only be one mother. Who is the mother, Mrs. Smith or Ms. Jones?
2. Under the applicable law of East Dakota, there can only be one woman who has the decision-making capacity regarding the continuation of the pregnancy. Who has that capacity in this case?
3. Is gestational surrogacy an ethically valid form of assisted reproduction?
4. Should the gestational surrogacy contract be legally enforceable?
5. Should this court grant the plaintiff's request for an injunction, thereby preventing Ms. Jones from having an abortion?

The reader is here invited to become a member of the ongoing jury evaluating reproductive technologies and to consider carefully these same questions.

The Case

Mr. and Mrs. Smith are an infertile couple. Mr. Smith has azopermia (inability to produce viable sperm) and is thus sterile. Mrs. Smith suffers from fallopian atresia (an inadequate development of the fallopian tubes). Her ovaries produce normal eggs and her endometrium is normal, but the eggs cannot get from the ovaries to the uterus. In their desire to have a child with some genetic linkage to at least one of them, they pursue in vitro fertilization with donor sperm. Eight viable embryos result. Four are implanted and the remaining four cryopreserved (frozen). The implantation yields a pregnancy with one child. The baby is healthy at cesarean section, but Mrs. Smith's uterus is damaged and has to be removed.

The Smiths feel an obligation to the cryopreserved embryos and believe that they should be given a chance to develop. They also believe that it is their responsibility to parent any resulting children because they created them in the first place. Consequently, they do not favor embryo donation, but decide to explore the use of a gestational carrier (i.e., surrogate mother — a woman who will carry the child to term for them).

After a diligent search throughout their home state of East Dakota, they contract with a woman, Ms. Jones, to serve as the gestational carrier for their embryos. As part of their contracting process they discuss who will have final decision-making authority in regard to various situations. Issues such as vaginal delivery versus cesarean section and the use of therapeutic abortion if the health of the carrier is jeopardized have been points of contention in similar relationships. They hope to avoid such controversies. In the contract the Smiths agree to let Ms. Jones have final decision-making authority with regard to medical questions concerning the health of the carrier and the mode of delivery.

At three months gestation an ultrasound examination of Ms. Jones reveals significant cardiac anomalies in the fetus. At this stage of development it is unclear what the final outcome will be, but there is a good chance that the child will be born with significant heart problems that will require surgical intervention.

Ms. Jones states that she wants an abortion. She believes that it is wrong to bring such a potentially ill and disabled child into the world and thereby subject the child to suffering. She indicates she does not want to be confined to another six months of pregnancy merely to bring into the world — if the fetus even survives — a child that she would abort if it were her own. She asserts that it is her right to abort under the contract.

Mr. Smith is torn and undecided, but Mrs. Smith clearly and strongly opposes any thought of abortion. She states that it is too early to make such a decision and that the child should be given as much of a chance as possible. She pleads with the carrier to continue the pregnancy; but Ms. Jones refuses to reconsider.

Mrs. Smith then appeals to the court for an injunction against the carrier to prevent her from having an abortion. Mrs. Smith contends that Ms. Jones does not have the right to make this decision or embark on this course because the issue is one that affects the health of the child only. The carrier's health is not affected in any way beyond the fact of the pregnancy itself — a situation she knowingly and voluntarily entered into — and nothing has changed in that regard. Therefore, the carrier cannot claim she was given this right in the contract, because the issue of maternal health and safety has not arisen. Since the Smiths will assume all care of the child at birth, the carrier incurs no additional expense or burden due to the child's potential disabilities. She therefore has no right to decide against the child's life.

Ms. Jones then contends that she is the one pregnant and therefore has a right to an abortion according to the U.S. Supreme Court case of *Roe v. Wade.* She claims that to continue the pregnancy would create a case of wrongful life.

146

Setting the Stage

C. Christopher Hook, M.D.

A brief comment about surrogacy may be helpful as we begin our discussion of this case (see Scott Rae's chapter in this volume for a fuller analysis). Surrogacy may occur in one of two ways. Traditional surrogacy is when a man's sperm is used to impregnate a woman, not his wife, with the intent of the surrogate surrendering the child to the genetic father and his wife at the time of delivery. Traditional surrogacy has been condemned because it forces a woman to surrender a child that is truly hers, not only in gestational terms, but also genetically. The other form of surrogacy is to use a gestational carrier. Here the child is conceived by in vitro fertilization, typically using the gametes of a husband and wife. In our case the egg of Mrs. Smith and donor sperm are joined to conceive a child. Accordingly, the woman who serves as the gestational carrier has no genetic linkage to the child. Gestational surrogacy has been more easily accepted by many than traditional surrogacy, because a carrier is not being asked to give up a child that is genetically hers.

Can the use of a gestational carrier be ethically justified in this case? The use of a gestational carrier could possibly be ethically justified if three criteria are met. The first two deal with the carrier: (1) the act must be truly a voluntary one, and (2) the carrier must be fully informed of the risks, the nature of her relationship with the commissioning couple, and the way that important treatment decisions will be made throughout the course of the pregnancy. If a woman truly understands the risks involved (and for this reason some would suggest that only women who have previously borne children should be allowed to be gestational carriers), gestational surrogacy may be seen as a true act of altruism. It involves giving the gift of life, and as such could possibly be acceptable. However, it is important that potential coercion be anticipated and avoided. Thus, many programs wisely forbid payment of the carrier, other than compensation for lost wages and for health care costs incurred in the course of the pregnancy. There should be no financial gain involved in this relationship. Because family members and friends are often considered or volunteer to serve as surrogates, it is also important to ensure there is no social coercion pressuring the individual to serve in this capacity.

The third requirement in order for this practice to be ethically acceptable deals with the medical indications for the use of the carrier. Only certain medical conditions should justify this process — for example, significant pathology of the uterus or uterine environment, which make successful pregnancy impossible. Congenital absence of the uterus, or (as in the case of Mrs. Smith) subsequent loss of the uterus with preexisting cryopreserved embryos are cases in point. The use of a gestational carrier should never be allowed for social reasons alone, such as when a woman wants to avoid the inconvenience of bearing her own child, or when the two people providing the sperm and egg are not married to each other.

The case at hand illustrates a significant concern: the fate of frozen (cryopreserved) embryos. By current standards of in vitro fertilization practice, many more embryos are often created than can be safely implanted at one time. Consequently, many couples opt to freeze embryos that are not initially implanted. Some implant these embryos at subsequent times, while others may not. The couple in this case simply cannot undergo subsequent implantations, even though that was their initial plan, because Mrs. Smith has lost her uterus. Her children are now in a frozen state of limbo and the only possible means by which they can continue to develop and enjoy life is through the use of a gestational carrier. It is the moral obligation of anyone who engages in in vitro fertilization to give all of the created embryos the fullest opportunity for a complete life. Amazingly, there are somewhere between 500,000 and 2 million cryopreserved embryos in the United States at this time — children in what Jerome Lejeune referred to as "The Concentration Can."[1] The moral acceptability of creating more embryos than can be implanted and the use of cryopreservation are critically important issues we must face in evaluating current assisted reproduction practices. Parents and providers of assisted reproductive technologies have a moral obligation to the lives that are conceived. For this reason, not only is gestational surrogacy ethically acceptable within the parameters defined above, but it may even be ethically obligatory as an act of rescue in certain circumstances.

Nevertheless, one must still recognize myriad potential pitfalls in the

---

1. Jerome Lejeune, *The Concentration Can* (San Francisco: Ignatius Press, 1992).

unique relationship between the gestational carrier and the commissioning couple. Fundamental questions regarding prenatal testing, use of therapeutic abortion, who has final decision-making authority over the mode of delivery, and the carrier's lifestyle during the course of the pregnancy must be resolved by all involved parties if the pregnancy and relationship are to be successful. As noted in this case, the Smiths and Ms. Jones had indeed discussed many of these issues and articulated them in a formal contract. Were the health of the mother to become threatened, they agreed, Ms. Jones would have the right to pursue a therapeutic abortion. They also left the choice of the use of cesarean section versus vaginal delivery in her hands because this would be an issue that would affect her health as well as the child's. However, they did not discuss what would happen if the problem that arose was with the child, not the carrier. The result was a significant disagreement between the parties.

There is little legal precedent governing this case. Most states do not formally recognize gestational surrogacy, nor would their laws enforce a gestational carrier contract. But an even larger legal question looms over this situation. Can a woman give up her "rights" as delineated in the case of *Roe v. Wade* in a gestational carrier relationship? The concern of the Court in *Roe* was to allow women reproductive liberty so that they would not have to incur the medical risks of pregnancies they do not desire or be required to bear children that they do not want to raise. The Smiths argue that Ms. Jones freely and knowingly entered into this relationship to bear a child for them. She will not bear any of the long-term responsibility for the care of the child, and therefore she has no interest beyond bringing the child safely to term. Her health status has not changed beyond the fact that she is pregnant — a condition she accepted when she entered the contract.

Ms. Jones argues that she is the one pregnant, that she is being affected emotionally by this pregnancy, and that it is still her explicit right to terminate the pregnancy at any time for any reason she chooses. However, if she chooses to terminate the life of the child, might she not in essence violate the procreative liberty of the Smiths?

One final and crucial question: Who is the genuine mother, that is, the one who should have the power over the child's future? Is she the genetic and intended social mother, or the mother who is bearing the child? As when a similar question was put to King Solomon long ago (see 1 Kings 3:23-27), wisdom may again point us to the woman who has the child's best interests at heart.

Rules and Exceptions

C. Ben Mitchell, Ph.D.

As the case before us illustrates, not only are the new reproductive technologies still under development, but so are the ethical and legal understandings of their appropriate use. Surrogacy represents only one area of burgeoning growth. Considering the various configurations and technologies, there are at least thirty-eight ways to "make a baby" today.

The last time I consulted the Internet on the topic of surrogacy, I found that icon of commercialization — an advertising banner — flashing back at me. The Genetics and IVF Institute has a cadre of "fully screened surrogates available for its patients," the banner said. In addition to rather pedestrian articles on the legal issues and updates on state laws governing surrogacy, I was able to peruse sites which included Creating Families, Inc., Surrogate Mothers Online, the National Association of Surrogate Mothers, and a special online gift shop for those involved in surrogacy, Expressions of the Heart.

Excluding the procreative relationship between a husband and a wife, surrogacy may be the oldest of the reproductive arrangements. There are two examples of surrogacy recorded in the Old Testament (Gen. 16:1-6; 30:1-13). This fact alone does not mean that we have a better-understood or more carefully explicated moral perspective on surrogacy than on other reproductive approaches. It means merely that we have lived with this possible arrangement longer than some of the others. With the advent of high-tech reproductive medicine, the increasing use of surrogates makes the necessity of critical moral reflection more urgent than ever. The now infamous case of Baby M was but a harbinger of what could become a legal and ethical quagmire. Sadly, the kinds of issues that arise in the case before us are not atypical.

The first thing to note about this case is that it was preventable. Mr. and

Mrs. Smith were so desirous of having children that they used in vitro fertilization (IVF). While some ethicists find this procedure morally appropriate, this case highlights at least one of the potential problems usually associated with IVF. In vitro fertilization often results in the fertilization of more eggs than a woman is willing or able to have implanted at one time. In this case, eight viable embryos were produced. Four were implanted in Mrs. Smith's uterus, and she carried one child to term. Four were frozen for later implantation. During delivery of the first child Mrs. Smith's uterus was damaged. She could not, therefore, have the other four embryos implanted.

Mr. and Mrs. Smith could not have foreknown that Mrs. Smith would have to undergo a hysterectomy following her first delivery. At the same time, with appropriate counseling the Smiths could have been informed of that possibility, remote as it may have been. Armed with that information they might have elected to have only the number of eggs fertilized that they were willing to have implanted the first time.

The reason some couples choose to have embryos frozen is because of the discomfort, health risks, and expense of retrieving and fertilizing eggs. Typically clinics will retrieve as many as a dozen eggs for fertilization. Four are usually implanted at a time. The remaining embryos are stored for later use. While the financial costs might have been greater, given the proper information, the Smiths might have elected to fertilize only the number of embryos they were willing to have implanted in one procedure. If the first procedure was successful or any serious damage was done to Mrs. Smith's reproductive system — both of which occurred in this case — they would not have had to worry about what to do with any remaining frozen embryos since none would have been produced. If, however, the procedure was unsuccessful they could have returned to the clinic to retrieve and fertilize four more eggs.

In addition to receiving information about the potential medical complications, the Smiths should also have been informed about possible social complications relevant to their decision. For example, they should have been asked to contemplate what they would do should either or both of them die between the time the embryos were produced and the time they would be implanted. In light of the *Davis v. Davis* case in Tennessee, they should have been encouraged to think about what would happen to the embryos should they divorce. These are not pleasant possibilities, but they should be broached in order to head off just the sort of problems the Smiths eventually faced.

The new reproductive technologies make the requirements of informed consent even more stringent than with some other procedures. Couples should be told of as many different potential outcomes as possible, within reason, so that they can preempt precisely these sorts of problems. With the advent of other technologies that make it easier to control how many embryos are produced, such as single-sperm injection, some couples might be able to avoid facing the trauma the Smiths had to endure. Couples need to be fully informed regarding all of their options.

The availability of assisted reproductive technologies raises another important issue for couples such as the Smiths; namely, what are their reasons for wanting children, and are those reasons sufficient warrant for using reproductive technologies?

Reasons prospective parents give for wanting children are legion. In his article, "Why Have Children?" Marshall Missner suggests that people choose to have children for either social or personal goals. They include:

Social goals
1. The survival of humanity.
2. The survival of one's culture or community.
3. Biological drive.

Personal goals
1. A simple desire to have children.
2. A "full" human life and young adulthood.
3. Financial benefit and/or improved social status.
4. Religious conviction.
5. A kind of personal immortality.
6. Enhancement of personal happiness.
7. Altruism.[1]

Are all of these reasons morally justifiable reasons for having children? Are all of these reasons for wanting children consistent with the risks involved in using assisted reproductive technologies? Do any of these reasons justify intentionally producing embryos whose fate is less than certain? These are important questions to ponder when considering reproductive technologies. Gilbert Meilaender's reflections earlier in the present volume are a good place to begin. Does any one reason, any combination of reasons, or the entire set of reasons together constitute a sufficient basis for embracing in vitro fertilization as a means of having a child? At the very least, couples should be encouraged to articulate their desire for children and balance those desires against the radical nature of reproductive technologies.

1. Marshall Missner, "Why Have Children?" *International Journal of Applied Philosophy* 3 (fall 1987).

In some ways, surrogacy arrangements are the most radical of reproductive practices. No other arrangement involves the gestational use of a third party. Egg and sperm donation certainly include important ethical issues of their own, but neither involves the use of another person's entire psychophysiology. Bringing a third party into the reproductive relationship between a couple is quite a risky proposition.

For instance, surrogacy arrangements challenge the traditional understanding of the nuclear family structure. In Western culture, the nuclear family has consisted of a mother, a father, and their children. The family is extended vertically through grandparents and horizontally through the marriage of children to spouses from different families. While adoption admittedly extends the family through the addition of children, this arrangement excludes other adults from entering the reproductive relationship. Moreover, however one construes the family, nearly all agree that familial relationships include moral obligations — specifically, the bundle of obligations revolving around the well being of the children.

Through surrogacy arrangements, whether traditional (where the surrogate provides the egg and is simply inseminated) or gestational (as here), an adult from outside the nuclear family enters the reproductive relationship. How and if this third party will interact with the nuclear family is not a question of minor import. If the surrogate is to be involved in the rearing of the child, this certainly expands the dynamics of the traditional family. What role will the surrogate play? Will she live in the home? Will she maintain regular visitation? Will she be responsible for the child's welfare? Will she be liable for the child should the child contract any illness or be involved in any future criminal activity? While various states treat these questions differently, none addresses the distinctively *moral* obligations of a surrogate. If she possesses no moral obligations, then she is treated merely as a bodily organ — a uterus.

In many cases both the family and the surrogate choose not to involve the surrogate in the life of the child post-natally. Is this arrangement ethical? Should not a person who plays a gestational role in childbirth have moral obligations to the offspring she brings into the world? We have learned through a very painful social experiment that biological fathers have moral, and sometimes legal, obligations to the children they sire, and we have imposed legitimately severe social sanctions on those who do not live up to those obligations. Parenthood carries moral obligations toward others. More analysis is needed of the moral obligations of electing to be a surrogate.

Also, unless the surrogate remains a part of the family unit, surrogacy forces a woman to dissociate herself from the baby developing within her womb. Even in cases of altruistic surrogacy, dissociation must take place. No one suggests that the act of giving up the child a surrogate has carried in her body for nine months is a "maternal" act. What, then, are we asking women to do in serving as surrogates? Sometimes surrogacy is compared with other "gift" arrangements such as blood or organ donation. In important ways, however, the analogy breaks down. Hardly any woman would say she feels the same psychological-emotional attachment to her blood or kidney she feels toward the child she is carrying. If her attachment to the child were very similar to her attachment to replenishable cells, we would not count that as something good. So, on what moral or even psychological grounds could we favor the level of dissociation necessary for a woman to serve as a surrogate? Even if we could find women who were able to separate themselves from the babies they are carrying, as Daniel Callahan has put it, "This is not a psychological trait we would want to foster, even in the name of altruism."[2]

Commercial surrogacy runs afoul of several considered moral judgments. For example, commercial surrogacy commodifies both the reproductive organs of surrogates and the children who are born through surrogacy arrangements. It is not coincidental that a colloquialism for surrogacy is "rent-a-womb." In its decision in the Baby M case, the New Jersey Supreme Court opined that,

> In a civilized society there are some things . . . that money cannot buy. . . . In America, we decided long ago that merely because conduct purchased by money was "voluntary" did not mean that it was good or beyond regulation and prohibition. . . . Employers can no longer buy labor at the lowest price they can bargain for, even though that labor is "voluntary" . . . or buy women's labor for less money than is paid to men for the same job . . . or purchase the agreement of children to perform oppressive labor . . . or purchase the agreement of workers to subject themselves to unsafe or unhealthful working conditions. . . . There are, in short, values that society deems more important than granting to wealth whatever it can buy, be it labor, love, or life.[3]

Payment for surrogacy makes marketable commodities of both a woman's uterus and the baby she carries. Commodification of humans treats them as means to another's ends and not ends in themselves. Their value is reduced to their value in economic exchange.

---

2. Daniel Callahan, "Surrogate Motherhood: A Bad Idea," *New York Times,* 20 January 1987, cited in Scott Rae, *Brave New Families: Biblical Ethics and Reproductive Technologies* (Grand Rapids: Baker Books, 1996), p. 163.

3. In *re Baby M,* Supreme Court of New Jersey, 537 A.2d, 1227, 1249 (1988).

Finally, the problem surrogacy is meant to solve is infertility. The experience of infertility is extremely traumatic for some couples. From whatever source and for whatever reasons, the desire to have children is incredibly powerful. We should not underestimate or ignore the psychological, emotional, and even spiritual impact of infertility among women and families. When the desire for children meets with any perceived "solution" there is a temptation to overlook the potential problems which may be associated with that solution. Physicians, nurses, clergy, and counselors should study carefully the impact of infertility upon people. Where possible, ethical reproductive alternatives to surrogacy should be encouraged. Where the causes of infertility remove all such alternatives, couples should receive compassionate counseling, including biblical counseling, so that they can understand that being childless does not make them less human or less able to serve their God. They also need the requisite information to seriously consider the potential problems associated with raising children who are not their own biologically.

The Smiths might have avoided the situation in the case before us by not allowing more embryos to be produced than they were willing to have implanted in one IVF attempt. But they did allow this excess. So the reality is that we have four embryos outside the womb. In cases like the Smith's, surrogacy may be justified as a form of rescue. An even better alternative, though, would be to avoid circumstances in which such rescue is necessary. Surrogacy should be the exception rather than the rule.

The Hidden Plaintiff

R. Martin Palmer, J.D.

The case before us may mislead us if it focuses our attention on the two women involved. More is at stake here than which woman's rights take precedence over the other's. Easily hidden from view is the one who has the most to lose in this situation — the one in the womb. How should we view this one so very small?

Life begins like everything else, at the beginning. At the moment of fertilization, a new human life begins. The human embryo is the youngest "baby picture" of each one of us; and to the eye of human beings, with our limited vision, we all look alike at this early stage. But the Creator sees our different faces and smiles already, although yet unformed, because he knew us before we were born and will knit us together in the temple of the womb.

The human embryo is a being; and being human, she is a human being. She is person and not property because no property has the property of building. Everything necessary to make the new human being — the entire blueprint to build a human brain capable of going to the moon and putting a foot on the moon — is there in the very beginning. Nothing is added after the moment of fertilization. It is all locked in. Not only the color of our hair and eyes, but even how long we will live, accident or sickness not intervening. The complete information necessary to build the new human being is written in the smallest subscript of the universe. We are fearfully and wonderfully made! The human embryo would fit on the tip of a pin. Yet, as prominent French geneticist Dr. Jerome Lejeune once observed, the embryo contains more information, perfectly organized, than would fit in one million NASA computers.

The human embryo just described, fitting on the tip of a pin, sounds very small. But what is big? What is little? To us, a professional basketball

156

player is big because he is taller than we are, and a baby in a crib is little because she is smaller than we are. But to God, what is big, what is little? To us the earth is big because it is much bigger than we are, and a grain of sand is small because it is much smaller than we are. But we are judging by the position of our own body and its size on the continuum of what is big and what is little. But to God, what is big and what is little? We look out beyond our big earth at night and see the infinite stars swirling in the expanses of space; and we now understand that in the grain of sand there are new universes of subatomic particles in "inner space" we have yet to discover (but which are already known to God). To God, what is big and what is little?

The late Dr. Lejeune reminded us that if we ask science: "What is a human being?" science responds: "A human being is a member of our species." And if we ask science: "What is a person?" science responds: "A person is a human being alive." And science has no other answer. Recently I ran across a paper which Dr. Lejeune gave me entitled "Is There a Natural Morality?" For an eleven-year period preceding his death, Dr. Lejeune and I worked closely together for the cause of children yet to be. I was greatly influenced by his wisdom and his gentle heart toward humanity. In this essay, Dr. Lejeune makes the important observation that, "The biological bomb is probably more dangerous for humanity than the thermonuclear bomb." The child in the womb, in the case before us, would take little convincing on this point — nor would countless millions of children threatened by intentional death before birth.

In the years immediately after he first made this statement, Dr. Lejeune had time to further refine his insight into the problem. He summed it up most succinctly in the last words of his final lecture series on American soil, a three-night lecture series ("Human Nature and Scientific Knowledge") he gave at Providence Hospital to the assembled academic and medical community of Washington, D.C., in October 1993:

> We have to rely on some touchstone, which we can faithfully employ to know what we should not do. Indeed, we have something that would tell us what is good and what is evil; and there is one phrase — and it is the only one — that everyone of us knows which could tell us without any technicalities, in every case what to do. If the politicians remember it, they can make honest laws; if the technicians do not forget it, technology will remain the honest servant of humanity. But, if both of them don't remember it, then you would have to deal with a denatured biology, and there would be a very dim future for mankind. This phrase, you know it, it judges everything and forever. It just says: "What you have done to the smallest of Mine, you have done it unto Me."

Merciful Compassion

Joyce A. Shelton, Ph.D.

The case sets Mrs. Smith, a mother unable to bear her own child, against Ms. Jones, the woman who has agreed to bear Mrs. Smith's child for her. Mrs. Smith's desire for a child has forced her to resort to in vitro fertilization. Her sense of responsibility for the lives of the extra fertilized embryos, her frozen attempts to have a child, have forced her to use a surrogate to rescue them. Ms. Jones, the substitute mother, becomes pregnant through embryo transfer. Now the two mothers learn from the doctors that the child may be born with severe cardiac abnormalities, which will certainly require surgical repair and may lead to a life of illness if not an early death. The scenario is in place for their clash.

The reactions of the two mothers are at stark odds with one another. Ms. Jones, who carries the child in her womb, pronounces that the merciful decision is death by abortion. Mrs. Smith, who carries the child in her heart, pleads that compassion dictates a decision for life. The dilemma, one of the lawyers in this case declares, is as old as the judgment of Solomon. Which of the mothers should prevail? The court is called upon to decide what the mothers cannot.

The shift of responsibility for a decision from the mothers to the court transfers the matter to a sterile, impersonal environment for dissection by law. The legal decision arrived at may satisfy a human judge, schooled in the tangled and often contradictory maze of judiciary procedure and precedent. However, the satisfaction of law does not necessarily entail the implementation of justice. A more important concern should be whether the decision is consistent with God's ethical standards. An analysis that goes beyond the legal, an examination of the motives of the heart, must also be made.

The case, to be sure, is fraught with underlying ethical issues. The purpose of the present essay is to focus on just one, the question that brings these mothers into emotional conflict. What is the best way to deal with the suffering child? Ms. Jones was willing to bring the child to term until she discovered that the child could be handicapped. She does not, she maintains, wish to be responsible for the child's suffering. The thought of bringing a suffering child into the world is distressful to her. It has affected her own mental and emotional health. Ms. Jones would bear no physical nor fiscal responsibility for the baby's care once the child is born, yet Ms. Jones feels compelled "to be merciful."

Although the ethical principle of "mercy" is invoked, further investigation is required before one accepts the principle's application in this case. To be merciful can mean to act in such a way as to relieve another person's pain and suffering. What Ms. Jones proposes, however, is a drastic measure, fetal euthanasia, or "mercy killing." In society's attempt to remove all reminders of unpleasantness or inconvenience, euthanasia is becoming increasingly acceptable. It is instructive to examine the dichotomies in society's methods of dealing with euthanasia at the beginning and at the end of life.

First, let us consider a decision to withhold treatment from a terminal patient. If the dying patient does not, in the estimation of the medical professionals, have the mental capacity to make the decision and has not left a directive, then a decision to withhold further treatment has traditionally fallen to the next of kin. Presumably, this is because a relative would have the best interests of the patient in mind. It would also be expected that the decision maker's mental faculties are not compromised. If the relative decides to withhold treatment with the express purpose of relieving the patient's suffering by hastening death, then the designation of euthanasia, in the sense of mercy killing, applies. The U.S. Supreme Court, while allowing the withdrawal of treatment so that the patient dies from the underlying disease, does not allow withdrawing treatment with the intent of causing death.[1]

Now let us turn to the abortion of a fetus with abnormalities — euthanasia of the pre-born child. In contrast to the current illegality of mercy killing at the end of life, the legal right to terminate the child's life in the womb has been given to the mother by the U.S. Supreme Court.[2] In light of the ethical guidelines for intervening in the care of a dying patient, one would expect that the mother would be required to have the child's best interests at heart and that her uncompromised mental capacity would be an essential prerequi-

1. *Vacco v. Quill,* 65 U.S.L.W. (1997).
2. *Roe v. Wade,* 410 U.S. 113 (1973).

site. Neither of these is required by law. In an odd twist of logic, the law allows the mother to choose death for her child when she can demonstrate that her own mental health is impaired by the continuance of the pregnancy. Thus, a legally sanctioned decision to abort a child is not predicated on the altruistic principle of mercy or even on the basis of undiminished mental capacity. Rather, the more self-serving purpose of relieving the mother's own suffering justifies the right of the mother to decree her own child's death. The irrationality of this law is glaring. That is why abortion proponents must justify the practice by proposing that a fetus, especially one that is "defective," is not a human being at all.[3] A legitimate concern is that soon those who are ill, dying, or handicapped, will also be widely seen as less than human beings.

Mercy is not the undergirding principle here. A mother may truly be concerned for the suffering of her unborn child and may wish to be merciful. Such a motive, however, cannot justify a decision to abort her child. To be merciful may indeed entail acting to relieve another's pain and suffering. Nevertheless, it is a quantum leap to assume that the person's death is a beneficial or even an acceptable means to that end. As Francis Beckwith explains: "It is amazingly presumptuous to say that certain human beings are better off not existing. For one thing, how can one compare existence with nonexistence when they do not have anything in common?"[4] Allowing such an extreme act imposes a value of pleasure, or at least the absence of pain, on life to such an extent that it not only results in the utilitarian removal of suffering, but of the sufferer as well. One would hope that even society's apparent preoccupation with self-gratification would not lead to acceptance of such a conclusion.

In choosing death for another, we presuppose that we can predict that person's future with certainty. Joseph Conrad once said, "to have his path made clear for him is the aspiration of every human being in our beclouded and tempestuous existence."[5] We can wish it, but we cannot have it. We are beings in time, not outside of time as God is, and we cannot see the future that God has planned for us or for another human being. Ms. Jones is convinced of her own merciful foresight in ruling that the child she bears should die. Yet, she bases her decision solely on data received from the medical community. It is true that sophisticated medical technology has given doctors great confidence that they can predict medical outcomes, but doctors are also

3. Francis J. Beckwith, *Politically Correct Death* (Grand Rapids: Baker Books, 1993), p. 65.
4. Ibid., p. 66.
5. Joseph Conrad, *The Mirror of the Sea* (Evanston, Ill.: Marlboro Press, 1988).

merely human beings. Patients can be misdiagnosed and the death sentences pronounced by doctors can be and frequently are overruled by God. Dr. William Cutrer has observed that sometimes we are all more sure than we are right.[6] Acknowledging our lack of foreknowledge, we should be far less hasty in our evaluation of either the significance or the consequence of another person's life.

Furthermore, appealing to mercy in order to justify depriving a child with a physical handicap of the chance for life suggests that perfect health is essential to a life of quality or worth. The value that society places on health is certainly affirmed by the popularity of exercise and fitness clubs as well as the huge number of dollars per capita spent on nutrition and medical care each year. Greater priority is given to health than to life itself. This emphasis on, indeed almost worship of, physical fitness leads to viewing a handicapped person as deficient, as lacking worth — that is, as being somewhat less than human. A further logical implication, in keeping with society's "disposable mindset," is that these "non-human" objects should be thrown away. It is easy, then, to lose patience and sympathy for the weak, the vulnerable, and the dependent. It becomes an all-too-simple solution to urge their departure in favor of persons who can actually contribute to the community.

To the Christian, whose beliefs are grounded in the premise that God created humanity in his image — that God is the giver and sustainer of life — this devaluation of some of God's image bearers on the basis of their lack of physical fitness cannot be defended. As John Kilner puts it:

> No individual life is beyond the scope of the life to which God is committed, no matter how unlovely or unworthy by human standards. . . . Some refer to the "value" of life, but this way of speaking improperly places life on a scale of value. It erroneously suggests that life can be compared and traded off with other things of value. Because life draws its significance from the realities of God and God's creation, however, it is not negotiable in this way. It is sacred, and we ought rightly to maintain a sense of awe or reverence about it. Even the notion of assigning "infinite value" to life misses the mark, for the issue is not how much value life has but the fact that its significance is not dependent on how valuable or useful it is to anyone.[7]

The quality of mercy is a virtue, but the affirmation of life is a principle that is given priority in the Bible. Christ says, "I came that they might have life, and might have it abundantly" (John 10:10b). We cannot express true

---

6. See his essay earlier in this volume.

7. John F. Kilner, *Life on the Line* (Grand Rapids: Eerdmans, 1992), p. 56.

mercy if, in so doing, we violate the principle of life. If one chooses to honor the principle of life, God honors the choice. He gives life, either here in this worldly existence, or in eternity. Deuteronomy 30:19 says: "I call heaven and earth to witness against you today, that I have set before you life and death, the blessing and the curse. So choose life in order that you may live, you and your descendants." To choose, instead, death, is to deny all that God, in his love for us, wishes us to have.

According to 1 Corinthians 15:26, death is an enemy. To assume that we have the right to choose death for another is a prideful act that stems from our desire to be in control. After all, there is hardly a greater control that one can have than control over the timing of a person's death. While it is difficult to watch someone we love suffer, and even more difficult to understand why God allows it, we have no right to assume the responsibility for ending someone else's existence just because we think we know better than God. We are adjured by God's Word to relinquish control to him by faith, because he created us, he loves us, and he alone has knowledge of our true best interests. "'I know the plans I have for you,' declares the Lord, 'plans for welfare and not for calamity, to give you a future and a hope'" (Jer. 29:11).

God has said that we are to be merciful, as he is merciful. If eliminating suffering by death is unacceptable, then what should our attitude be toward the sufferer and suffering? Our proper response to the sufferer is compassion. Compassion means "to suffer with," to share in the sufferings of the sufferer. God has promised that he will be with sufferers in their need. His provision for the sufferer is grace to endure. The difficulty for the one who watches another suffer, is that, while God gives grace to the one who must endure, he does not always give that grace to the one who must watch. True compassion helps us to be brave and even helpful watchers. The value of shared suffering is that it tests faith, it strengthens bonds of relationship, it solidifies commitment, and it brings an admission that we are not and cannot be in control. Our perspective of who we are (the servant) and who God is (the Lord) becomes more real. As John Kilner says, "The more people recognize that life and ethics are about character and virtue as well as right decision making, the more they will appreciate the significance of the contributions suffering can make to their lives."[8] Compassion helps us learn to love by placing our focus on another person's needs rather than on our own.

The true nature of suffering — suffering from God's perspective — is an enigma to us. Elisabeth Elliot writes: "Each time the mystery of suffering touches us personally and all the cosmic questions arise afresh in our minds,

8. Ibid., p. 104.

we face the choice between faith and unbelief. There is only one faculty by which we may lay hold of this mystery. It is the faculty of faith, and faith is the fulcrum of moral and spiritual balance."[9] As hard as it is to understand, God may use suffering for our good and for the good of others. We all suffer or have suffered in some form or another; yet, in our never-ending pursuit of pleasure and avoidance of pain, we do not see suffering as having value. God does. Many Scriptures deal with suffering, but one that is difficult to evade is one that celebrates knowing "Christ and the power of his resurrection, and the fellowship of his sufferings, being conformed to his death" (Phil. 3:10). Somehow, for some reason, suffering is an unavoidable part of knowing Christ. Christ himself did not escape it. If we try, by engineering our own death or that of another, the implication is that we will leave this world without understanding a dimension of Christ that we need to know. Perhaps the answer has to do with what Christ meant when he said "he who does not take his cross and follow me, is not worthy of me" (Matt. 10:38). The cross is an implement of suffering. Our eternal life was bought at the price of great suffering. We need to learn that neither our own life, nor any one else's life, belongs to us.

These concepts of mercy and compassion speak directly to the case at hand. The lawyers and judges may ultimately decide that Ms. Jones has the legal right to exercise her personal sense of mercy, however misguided, and abort her unborn child. The "judgment of Solomon," however, must be on the side of Mrs. Smith. She is the compassionate mother who refuses to believe that her handicapped child's life has less worth in the eyes of God than any other child who was created by him and whose life was preciously paid for through the suffering and death of Jesus Christ.

9. Elisabeth Elliot, *A Path Through Suffering* (Ann Arbor, Mich.: Servant Publications, 1990), p. 98.

The Unchosen and Frozen

Samuel B. Casey, J.D.

The case before us provides a vivid illustration of many of the ethical problems surrounding in vitro fertilization (IVF) that have already been discussed in the present volume's chapters by Dennis Hollinger and Teresa Iglesias. In order to develop a life-affirming and constitutionally sound way of handling such legal cases, this essay will examine how the law to date has treated them. The essay will conclude with a constructive proposal.

Despite the high costs, practical uncertainties, low success rates, and physical burdens of in vitro fertilization, a growing number of infertile couples who cannot naturally conceive offspring are turning to this technology.[1] Because fertility drugs given to women in IVF programs sometimes produce more embryos than can safely be implanted at any one time, it has become quite common to freeze the unused embryos, for later thawing and implantation, via a process known as cryopreservation.[2] Although cryopreservation may relieve the woman of the cost and physical burden of further egg retriev-

---

1. Janette M. Puskar, "Prenatal Adoption: The Vatican's Proposal to the In Vitro Fertilization Disposition Dilemma," *N.Y. Law School Journal of Human Rights* 14 (1998): 757. IVF costs are reported to range between $8,000 and $10,000 for a cycle of ovum retrieval and implantation, and up to $3,500 in medication costs per cycle. See Bette Harrison, "Special Delivery: A Baby," *Atlanta Journal,* 21 December 1997; Abigail Trafford, "Medicine's Money Back Warranty," *Washington Post,* 5 August 1997, Health section, p. 6. The couple in the *Davis v. Davis* case spent $35,000 on IVF; the couple in *Kass v. Kass* spent over $75,000. David L. Theyssen, "Balancing Interests in Frozen Embryo Disputes: Is Adoption Really a Reasonable Alternative?" *Indiana Law Journal* 74 (1999): 711, 734 n. 221.

2. See Clifton Perry and L. Kristen Schneider, "Cryopreserved Embryos: Who Shall Decide Their Fate?" *L. Legal Med.* 13 (1992): 463, 468.

als while temporarily preserving the lives of the frozen embryos, it also brings along with it troubling medical, ethical, and legal issues. Difficulties multiply when the parties divorce, die, or disagree about the disposition of the frozen embryos during the lengthy period that cryopreservation permits between fertilization and implantation.[3]

There are more than 10 million infertile couples in the United States. In the last decade the infertility industry has grown from about 30 to over 300 clinics earning revenues in excess of 1 billion dollars. It is estimated that in the United States in 1999 more than 75,000 infants will be born as a result of IVF — more than twice as many as will be available through traditional adoption. At the same time, hundreds of thousands of human embryos are now frozen, suspended in liquid nitrogen tanks (with more than 19,000 more frozen embryos estimated to be added each year).[4] Yet while there are pages of adoption laws on the books in each state, only nine states — Louisiana, New Mexico, Florida, Pennsylvania, Kentucky, Kansas, Virginia, New Hampshire, and California — have enacted some legislation relating to IVF, and only three of these states — Louisiana, New Mexico, and Florida — regulate the disposition of the embryo in any way. Consequently, in most states the fate of most frozen embryos is left open-ended and undecided.[5]

---

3. See Knoppers and LeBris, "Recent Advances in Medically Assisted Conception: Legal, Ethical, and Social Issues," *Am. J. L. & Med.* 4 (1991): 329. Without cryopreservation, all fertilized eggs must be implanted immediately or these human embryos will expire. See Marcia Joy Wurmbrand, "Frozen Embryos, Moral, Social, and Legal Implications," *S. Cal. L. Rev.* 59 (1986): 1079, 1083.

4. Lori B. Andrews, "Embryonic Confusion," *Washington Post,* 2 May 1999, pp. B1, B4. "Not all embryos survive the freeze-thaw process. A 50% survival rate is considered reasonable. After the thaw, embryos retaining 50% or more of the cells they had before freezing are cultured and placed back in the uterus via a tube inserted in the cervix. The number returned varies with the desires of the patient under the guidelines of age categories; under 35 years old, up to four embryos, 35 years and older, up to six embryos. National statistics for women 39 or less is 27% per embryo transfer, for women over 39, 14% per embryo transfer. Delivery rates will be lower due to miscarriage." IVF Phoenix Infertility Information Booklet, *http://www.ihr.com/fertbook/treatment.htm.*

5. La. Rev. Stat. Ann §§9:123-133; N.M. Stat. Ann. §24-9A-1 to 7; Fla. Stat. Ann. §§742.11 et seq.; 18 Pa. Const. Stat. Ann. §316(c)(1989); Ky. Rev. Stat. Ann. §311.715; Kan. Stat. Ann. §65-6702; N.H. Stats., Title XII, Chapter 168-B; Va. St.§20-156 et seq. See generally Judith F. Daar, "Regulating Reproductive Technologies: Panacea or Paper Tiger," *Hous. L. Rev.* 34 (1997): 609.

The Louisiana law, passed in 1986, declares that an in vitro fertilized ovum is a "juridical person" that can be disposed of only through implantation. If IVF patients "fail to express their identity" or renounce parental rights, the physician is deemed to be the embryo's temporary guardian until adoptive implantations can occur. La. Rev. Stat. Ann. §9:123, §9:123. In

The only federal law related to artificial reproductive technologies such as IVF is the Fertility Clinic Success Rate and Certification Act of 1992.[6] It

---

Louisiana, then, an agreement may not direct embryo dispositions other than implantation. Other options, such as donation for research or destruction, are prohibited. The federal constitutionality of this provision has not yet been tested.

The New Mexico statute prohibits all research involving embryos and fetuses, including IVF research. The statute allows the use of IVF to treat infertility as long as all the embryos are transferred to human recipients. N.M. Stat. Ann. §24-9A-1 to 7 (Michie 1997).

The Florida law is the only other law addressing prior embryo disposition agreements. Florida's law provides that couples undergoing IVF must enter into a written agreement that provides for the disposition of embryos in the event of death, divorce, or other "unforeseen circumstances"; and in the absence of a written agreement, the gamete providers are given joint dispositional authority. Fla. Stat. Ann. §742.17. The attempt in the first portion of this law to avert dispositional disputes by requiring prior agreements is effectively thwarted by the latter provision stating that in the absence of an agreement, the disagreeing gamete providers have joint authority — an invitation for judicial involvement without any legislative guidance.

The Pennsylvania statute requires all fertility clinics to file information concerning the location and staffing of each clinic, as well as the number of eggs fertilized, implanted, and destroyed. 18 Pa. Const. Stat. Ann. §3213(e). The Kentucky statute bars the use of public funds for the purpose of "obtaining an abortion" or the "intentional destruction of a human embryo." Ky. Rev. Stat. Ann. §311.715. The Kansas statute authorizes human embryo destruction by prohibiting any limitation on "the use of any drug or device that inhibits or prevents ovulation, fertilization, or implantation of an embryo and disposition of the product of in vitro fertilization prior to implantation." Kan. Stat. Ann. §65-6702. The Virginia statute requires all fertility clinics to obtain a disclosure form executed by the patients indicating their knowledge of clinic practices, success rates, and other matters. Va. Code Ann. §54.1-2971.1 (Michie 1997). The New Hampshire statute restricts access to IVF and embryo transfers to women over twenty-one who meet stipulated medical criteria and requires a medical evaluation of all gamete providers. N.H. Rev. Stat. Ann. §168-B:13. The California laws make it "unlawful for anyone to knowingly use sperm, ova, or embryos in assisted reproduction technology, for any purpose other than that indicated by the sperm, ova, or embryo provider's signature on a written consent form" (Cal. Penal Code §367g [West 1998]); and require the physician and surgeon who remove sperm or ova from a patient "to obtain the written consent of the patient" before the "sperm or ova are used for a purpose other than reimplantation in the same patient or implantation in the spouse of the patient." Cal. Bus. & Prof. Code §2260 (West 1998).

6. 42 U.S.C. §§201, 263a-1 to 263a-7 (Supp. 1996). The act requires the Centers for Disease Control to promulgate model certification procedures for fertility clinics and laboratories "to assure consistent performance of artificial reproductive technology procedures, quality assurance, and adequate record keeping at each certified embryo laboratory," as well as to standardize the reporting of pregnancy success rates. The reforms are minimal since the information required is otherwise available, neither laboratory certification nor clinic reporting are required by the act, and the only "penalty" for failure to

neither governs nor regulates the disposition of frozen embryos. Nor are there any federal court cases establishing federal law on the subject of frozen embryos.

Other countries have addressed this dilemma in different ways. Germany avoids the problem entirely by prohibiting the freezing of embryos.[7] Great Britain instituted a law mandating the destruction of any frozen embryos that have been in a cryopreserved state for over five years.[8] In 1996, appalled by the imminent mass destruction of thousands of frozen embryos, the Vatican responded by suggesting "that married women volunteer to bring the embryos to term in 'prenatal adoption.'" At the time, there were approximately 9,300 fertilized embryos subject to destruction under the law. Of these 9,300 embryos, the clinics were able to reach the "parents" of 6,000 of the embryos, who exercised their rights under the law either to extend the storage for another five years or to donate the embryo to research or to another woman. The 3,300 remaining "potential infants" — in a chilling reminder of another Holocaust involving persons considered by their Nazi executioners to be "sub-human" — were thawed out, destroyed with saline solution, and incinerated with other biological waste.[9]

---

comply with the act's recommendations is public identification as a program that has failed to do so. See Donna A. Katz, "My Egg, Your Sperm, Whose Preembryo? A Proposal for Deciding Which Party Receives Custody of Frozen Embryos," *Va. J. Soc. Pol'y & L.* 5 (1998): 623, 630.

7. Ibid., p. 776, n. 28. Another way of avoiding the problem entirely would be the development of a reliable means of freezing and thawing unfertilized eggs, since a reliable means already exists for freezing and thawing sperm. Because the potential harm of freezing would occur before conception, and thus before the beginning of life, "right to life" concerns would not be implicated. However, the current technology for freezing and thawing unfertilized eggs is extremely unreliable. Furthermore, even after the technology develops to reliably freeze and thaw individual eggs, there will still be thousands of previously frozen embryos whose fate will have yet to be determined. Indeed, one commentator estimates that as many as 20,000 frozen embryos across the country may be the subject of custody disputes. See Katz, "My Egg," p. 674 (n. 79).

8. Human Fertilisation and Embryology Act, 1990, ch. 37 §14 (Eng.). See generally Puskar, "Prenatal Adoption."

9. James Walsh, "A Bitter Embryo Imbroglio: Amid Dramatic Protests and Universal Unease, Britain Begins Destroying 3,300 Human Embryos," *Time,* August 1996, p. 10. The idea of "prenatal adoption" is neither new nor bizarre, as some IVF clinics require couples either to implant the frozen embryo in the natural mother or donate it for implantation in another woman. For example, Louisiana and New Mexico law requires mandatory donation of the unused frozen embryo for purposes of implantation only, not destruction. While the constitutionality of a statute mandating "prenatal adoption" has not yet been determined, at least one commentator has concluded that such a statute would not infringe upon the federal privacy right because the "possibility of an anonymous hereditary

In Australia, the Infertility Treatment Act of 1984 establishes time restrictions similar to those in Britain, with the possibility of extension for reasonable grounds. The act requires that when implantation in the woman donor is not possible, the embryo should be made available to another woman with donor consent. The embryos can be destroyed if donors do not consent or withdraw their consent in writing. However, when a donor cannot be contacted, the government has the authority to order the hospital where the embryo is stored to implant the embryo in another woman.[10]

Since 1978, when the first IVF baby, Louise Brown, was born in Great Britain, U.S. courts have also decided issues related to frozen human embryos in cases such as *Del Zios v. Columbia Presbyterian Medical Center*,[11] *York v. Jones*,[12] *Davis v. Davis*[13] and *Kass v. Kass*.[14] In addition to these (mostly appel-

---

tie" is at best only a psychological, not a physical, burden and should therefore not be deemed an "undue burden" outweighing the state's interest in the potential life being protected by a mandatory prenatal adoption statute. Puskar, "Prenatal Adoption," pp. 762-63.

10. See generally Heidi Forster, "The Legal and Ethical Debate Surrounding the Storage and Destruction of Frozen Human Embryos: A Reaction to the Mass Disposal in Britain and the Lack of Law in the United States," *Wash. U. L. Q.* 76 (1998): 759, 761-64.

11. *Del Zios* was the first unreported case involving frozen embryos. The Del Zioses filed suit when a supervising doctor at Columbia Presbyterian Medical Center intentionally destroyed their human embryo, obtained at the hospital through IVF, because the doctor felt IVF constituted an unauthorized and "unwarranted practice which posed danger to any human life resulting from such experimentation." The jury rejected the Del Zioses' conversion claim for an unexplained reason, but returned a verdict for $50,000 for intentional infliction of emotional distress.

12. The dispute in *York* concerned one frozen fertilized embryo that the plaintiffs, Mr. and Mrs. York, sought to have released and transferred from an IVF clinic in Norfolk, Virginia, to another clinic in Los Angeles. 717 F. Supp. 421 (E.D.Va. 1989). The clinic refused to transfer the embryo and the Yorks sued for possession. The Yorks won when the court implicitly adopted the "embryo as property theory" by noting that a bailor-bailee relationship was created. The cryopreservation agreement between the Yorks and the clinic created a bailment because the clinic had legal possession of the embryo and had the duty to account for it. The court stated that when the purposes of the bailment relationship have terminated, "there is an absolute obligation to return the subject matter of the bailment to the bailor." The court found that the agreement between the parties recognized the Yorks' property rights and limited the clinic's rights as bailee. The court denied the clinic's motion to dismiss, and eventually the case settled. See generally Bill Davidoff, "Frozen Embryos: A Need for Thawing in the Legislative Process," *SMU Law Review* 47 (1993): 131.

13. *Davis* is the leading case, which made its way up to the Tennessee Supreme Court. There the court decided the fate of seven frozen embryos. 842 S.W.2d 588 (Tenn. 1992). A divorced couple, Mary Sue Davis and Junior Davis, were arguing over the disposition of their cryogenically preserved embryos resulting from IVF procedures. Mrs. Davis

late) court decisions, there are at least three trial court decisions, two presently pending on appeal, that suggest the need for state legislative action. Such action is sorely needed, if for no other reason than to avoid the increasing repetition of costly litigation, unpredictable results, unacceptable delays,

---

wanted to donate the remaining frozen embryos to another infertile couple, while Mr. Davis wanted them destroyed. The trial court, W. Dale Young presiding, based upon the expert testimony of the Nobel Prize–winning geneticist, Dr. Jerome Lejeune, held that the frozen embryos were "human beings" and awarded "custody" to Mrs. Davis so that she could have them implanted. At that time, Mrs. Davis wanted the frozen embryos to be implanted in her, and Mr. Davis wanted the embryos to remain frozen while he decided what to do. However, between 1989 when the case was decided by the trial court and 1992 when the Supreme Court of Tennessee rendered its decision, both parties had remarried and changed their positions.

The court of appeals reversed the trial court's finding that frozen embryos were human beings, likening them more to personal property, and concluded that Mr. Davis had a "constitutionally protected right not to beget a child where no pregnancy has taken place" and that "there is no compelling state interest to justify ordering implantation against the will of either party." The court of appeals remanded the case to the trial court for entry of an order giving both parties "joint control and equal voice over their disposition" because both parties shared an interest in their frozen embryos."

The Supreme Court of Tennessee stated that although the Davises had an interest in the nature of ownership and disposition of the embryos, the interest was not a true property interest. The court concluded that frozen fertilized embryos deserve "special respect because of their potential for human life." The court also noted that the essential issue was whether the parties would become parents, not whether frozen embryos are "property" or "persons." The court concluded that the answer to this dilemma "turns on the parties' exercise of their constitutional right to privacy." The court discussed the cases where the Supreme Court held that the right to procreate was "one of the basic civil rights of man" (citing *Skinner v. Oklahoma,* 316 U.S. 535, 541 [1942]). The court also noted that "the right of procreational autonomy is composed of two rights of limited but equal significance — the right to procreate and the right to avoid procreation." The court held that in disputes concerning the disposition of a frozen embryo, a court should look first at "preferences of the progenitors," and if there is a dispute, then a court should look at the prior agreement pertaining to the disposition of the embryo. If, in a case such as this one, there is no prior agreement, the court stated that it must resolve the dispute over "constitutional imports" and weigh the interests of Mrs. Davis in her right to procreate and Mr. Davis in his right not to procreate. The court noted that "the party wishing to avoid procreation should prevail" if the other party has other reasonable means of procreating. If that party has no reasonable means of procreating, then the court will consider their argument for using the frozen embryo. In this case, Mrs. Davis wanted to donate her embryos to another couple, and the court held that in such a case, "the objecting party obviously has the greater interest and should prevail." Consequently, the Tennessee Supreme Court ruled that the Knoxville Fertility Clinic could destroy the Davis's human embryos.

14. The dispute in *Kass* involved the issue of the disposition of five frozen fertilized embryos. 663 N.Y.S. 2d 581 (App. Div. 1997). The parties in this case — now divorced —

public intrusion in private lives, and ongoing familial disruption.[15] At present, the courts have tended to side with the party seeking the destruction of human embryos. Judges have either cited contractual grounds or held that the interest in avoiding procreation trumps all other asserted interests, in-

---

were arguing over their frozen embryos. However, unlike the situation in the *Davis* case, when the Kasses underwent IVF treatments, they signed an informed consent document which provided in pertinent part: "In the event that we no longer wish to initiate a pregnancy or are unable to make a decision regarding the disposition of our stored, frozen pre-zygotes, we now indicate our desire for the disposition of our pre-zygotes . . . (b) Our frozen pre-zygotes may be examined by the IVF Program for biological studies and be disposed of by the IVF Program for approved research investigation as determined by the IVF Program." During the divorce proceedings, the parties executed a document that set forth their understanding of what was previously agreed to in the informed consent document as to the disposition of the remaining frozen embryos. However, Mrs. Kass had a change of heart and wanted to have the embryos implanted in *her* uterus. Mr. Kass wanted to have them turned over for embryo research, as provided for in the informed consent document.

The trial court awarded Mrs. Kass the five embryos for implantation. The trial court first reasoned that embryos enjoy the status of something between human life and property. The court stated that, just as in vivo fertilization, a father's right ends at the moment of fertilization, and thus the issue of embryo disposition is a matter exclusively within the mother's discretion. The court further found that the informed consent document agreed to by the parties was not dispositive and agreed that in the event of a divorce, a court would determine the issue of disposition. Therefore, the trial court held that Mrs. Kass was to determine the fate of the five frozen embryos.

The appellate division reversed and held that the informed consent document and the uncontested divorce instrument clearly stated the parties' intent that in the event of a divorce, any remaining embryos were to be donated for scientific research. The court first criticized the lower court's analysis and stated that it had committed a "fundamental error." The court stated that the Supreme Court was wrong to "[equate] a prospective mother's decision whether to undergo implantation of pre-zygotes which are the product of her participation in an IVF procedure with a pregnant woman's right to exercise exclusive control over the fate of her non-viable fetus."

Rather, the court found that the first inquiry should be whether the parties had made an "expression of mutual intent which governs the disposition of the pre-zygotes under the circumstances in which the parties find themselves." The court relied on the *Davis* decision where the Supreme Court of Tennessee stated that a critical factor in determining the disposition of frozen embryos is whether there was a prior written agreement. The *Kass* court found that because the Kasses executed the informed consent document, which was clear and unambiguous, the embryos would have to be disposed of as provided for in that agreement. The court also noted that Mrs. Kass did not meet the evidentiary standards to permit the balancing test, as used in *Davis v. Davis*, which balances the parties' interests in using the embryos. The court concluded that "the decision to attempt to have children through IVF procedures and the determination of the fate of cryopreserved pre-zygotes resulting therefrom are intensely personal and essentially private matters which are appropriately resolved by the prospective parents rather than the courts." The

cluding the state's interest in protecting human life and the parties' contractual rights.[16]

The cases essentially turn on two questions: (1) what is the legal status of the frozen embryo; and (2) ought this status be determined by the gamete

---

court determined that if the parties had entered into a prior agreement indicating their mutual intent regarding the disposition of any unused frozen embryos, then a court must not interfere. Therefore, because the Kasses had executed the informed consent document, the court held that the Kass embryos be retained by the IVF program for scientific research.

On May 7, 1998, the appellate division's decision was affirmed on appeal to the New York Court of Appeals, where the court held that the parties' agreement providing for donation to the IVF program controls the issue. 696 N.E. 2d 174 (N.Y. 1998).

15. The three unreported trial court frozen embryo cases are the Massachusetts case of *AZ v. BZ,* the New Jersey case of *J.B. v. M.B.,* and the Michigan case of *Bohn v. Ann Arbor Reproductive Medicine Associates.* In *AZ v. BZ* (Suffolk County Probate Court, March 25, 1996), despite seven separate express contractual agreements which, in the event of divorce or separation, gave the wife custody of the frozen embryos for implantation in her uterus, the court held that "the party who wishes to avoid parenthood should prevail when weighing the relative interests, so long as the other party has a 'reasonable possibility' of achieving parenthood through other means (like further IVF or adoption)." The court supported its holding by noting its interest in ensuring that the children receive the love and support of two parents, in avoiding encumbering the unwilling parent with an unwanted financial burden which it appeared Massachusetts law would otherwise impose, and in giving precedence to the negative right not to procreate because it better eliminates burdens on both the unwilling parent and potential child. With respect to the frozen embryos' "right to life," the *AZ* court diminished it by its current non-viability and extremely low probability of further development. The *AZ v. BZ* case settled without an appeal. See David L. Theyssen, "Balancing Interests in Frozen Embryo Disputes: Is Adoption Really a Reasonable Alternative," *Ind. L.J.* 74 (1999): 711, 732 (cogently arguing that the availability of adoption should not be deemed to meet the "reasonable possibility" of parenthood standard articulated in *AZ v. BZ*).

In the New Jersey case of *J.B. v. M.B.* there was no written agreement, and it was the Catholic husband who sought to bar his former wife from destroying the frozen embryos they produced while they were still married. The court rejected his claim to assume all parental responsibilities in order to preserve the embryos to implant in another woman or donate them to an infertile couple because "he is quite healthy and fully able to father a child. In such a case [like *Davis*] where the party seeking control of the frozen embryos intends merely to donate them to another couple, the objecting party has the greater interest and should prevail."

The *Bohn* case in Michigan, currently pending on appeal, involves a suit against a fertility clinic for custody of three frozen embryos, brought by the woman who conceived them by IVF while she was married to her former husband. The woman sought to have the embryos implanted in her as provided in her alleged written contract with the fertility clinic, which was signed only by her, not her former husband. The divorced husband intervened in the case arguing that the implantation of the embryos in his former wife violates

171

providers as a matter of private contractual law or by the state as a matter of public policy, particularly when the gamete providers either disagree, die, divorce, or are otherwise unavailable. Regarding the status of the frozen embryo, there are four basic theories: (1) the "human life" theory that the frozen embryo is a human being deserving the equal protection of the laws against homicide; (2) the "pure property" theory that the frozen embryo is merely property to be dealt with under common law contract and personal property principles; (3) the "special status" theory that the frozen embryo deserves special respect as against the claims of the non-gamete providers, but does not possess rights sufficient to limit the procreative or contractual rights of the gamete providers themselves; and (4) the "constitutional rights" theory, based upon analogies to distinguishable Supreme Court right to privacy decisions protecting abortion and contraception. This last theory holds that the unborn have no constitutionally protectable right to life, but that a gamete provider does have a right not to procreate. This procreational right trumps any constitutional, statutory, or contractual rights that the state or any other gamete provider may assert against the destruction of a human embryo prior to implantation. Each of these four theories has its proponent and critics.[17]

---

his "[constitutional] right not to procreate and that there was no agreement between himself and his former wife giving her exclusive custody over the remaining embryos to do with them as she in her sole discretion may decide. The wife seeks to protect the lives of the embryos while protecting her right to affirmatively procreate, which under the facts of this case she argues can most likely be accomplished through the implantation of the remaining three embryos in her uterus." The court held that "neither [the former wife or the former husband] have a unilateral contractual right to take possession of the embryos, and until such time as they agree on a disposition, the embryos may remain [in their cryopreserved state]." The *Bohn* court further held that the Michigan Child Custody Act, by its terms and intent, does not apply to frozen embryos. Finally, the court held that the objecting gamete provider "has a constitutional right not to beget offspring, and, therefore, has the right to veto any use of the embryos to produce more children."

16. The American Bar Association, at its 1998 meeting in Nashville, Tennessee, considered adopting a policy statement, proposed by the Family Law Section (Report No. 106), about frozen embryo disposition. The ABA indefinitely postponed the policy. The proposed ABA policy states that it is intended for use "in cases of marriage dissolution where the couple has previously stored frozen embryos with the intent to procreate." The policy suggests that if the marriage has dissolved, the couple is in disagreement about the fate of the embryos, and there is no pre-procedural agreement, "the party wishing to proceed in good faith and in a reasonable time, with gestation to term, and to assume parental rights and responsibilities should have possession and control of all the frozen embryos."

17. There are more than 320 law review articles on the legal controversies surrounding frozen human embryos. See Puskar, "Prenatal Adoption," pp. 765-66; Katz, "My Egg," pp. 634-39. "The [Constitutional] theory, . . . is distinct from, but in accordance with the

Those who hold that the embryo is a living human being seek protective legislation like the laws of Louisiana and New Mexico permitting "prenatal adoption" of frozen embryos and prohibiting the intentional destruction of human embryos. Such proponents also support judicial application of the "best interests of the child" standard for determining which of the gamete providers ought to have custody of the frozen embryos, regardless of any written agreements the gamete providers may have to destroy or donate the embryos for destructive research. Based upon Dr. Jerome Lejeune's expert genetics testimony — calling human embryos "early human beings" and "tiny persons," and equating the destruction of a frozen embryo with death from a "concentration can" — the trial court in the *Davis v. Davis* case adopted the human life theory. The court gave custody to the mother who wished to preserve the lives of her frozen embryos, rather than to the husband who desired to destroy them because the couple had subsequently been divorced.

Those who hold that frozen embryos are pure property generally support legislation that (a) affirms the gamete providers' joint decisional authority over their frozen embryos, with or without a pre-conception agreement; (b) requires them to enter an agreement providing for the broad range of dispositional alternatives, including destruction, storage, donation for research, and donation to another infertile couple; and (c) ensures the enforce-

---

[special status theory]. . . . Because the Supreme Court [denied certiorari in *Davis v. Davis*, 842 S.W.2d 588 (Tenn. 1992), cert. denied 507 U.S. 911 (1993)] has not decided a case involving the custody of [human embryos], the statement of its view regarding their status is based upon the extrapolation from the Court's analysis of the status of the unborn. In *Roe v. Wade*, 410 U.S. 113 (1973), the Court held that the word 'person' 'has application only post-natally . . . [and,] as used in the Fourteenth Amendment, does not include the unborn.' Logically then, if the word 'person' does not apply to the fetus, which is at a considerably more advanced stage of development, it does not apply to a human embryo that is not even located in the uterus."

On the other hand, the constitutional theory is limited in its application to human embryos *ex utero* because *Roe* has been held to be limited to abortion, and, "outside the context of abortion, the states can protect the unborn, human being at every stage of development." See Clarke D. Forsythe, "Human Cloning and the Constitution," *Valparaiso U. L. Rev.*, vol. 32, pp. 469, 500-502, n. 141 and accompanying text. "Recently, Indiana became the twenty-sixth state to treat as a homicide the killing of an unborn human being as some stage of gestation when it enacted a law, over the Governor's veto, to treat the killing of an unborn child as homicide whether the child was born alive or not. In addition, Michigan and Wisconsin enacted legislation in 1998 to protect the unborn child ('embryo' and 'fetus') at all stages of gestation. Thus, states continue to extend legal protection to the unborn human being throughout gestation, and, outside the context of abortion, a remarkable legal and legislative consensus exist across at least thirty-eight states that the life of a human being begins at fertilization or conception."

ability of such prior directives. In the absence of prior agreement, pure property proponents generally assert that courts should prohibit the use of frozen embryos by either party without the consent of the other. Such a theory was the approach taken by the Tennessee Court of Appeals in *Davis v. Davis* in reversing the trial court decision based upon the human life theory.

Those who hold the special status theory take an intermediate view between the pure property theory and the human life theory. The frozen embryo is not given the full status afforded a human being, but is not considered pure property either. According to this theory, the embryo warrants greater respect than human tissue because of its potential life. Proponents of this theory respect the embryo as a "symbol of human life." The American Fertility Society has adopted this position. The special status theory was the approach adopted by the Tennessee Supreme Court in affirming the result, if not the reasoning, of the Tennessee Court of Appeals in *Davis v. Davis*.

Based on the unpredictable, inconsistent, and unduly costly results described above, courts and commentators appear to agree that the United States, on the federal and state levels, should adopt a legislative framework regulating the cryogenic preservation and subsequent implantation of human embryos. Unfortunately, many of the proposed frameworks ignore the humanity of the human embryo by recommending that they be likened to property, not people, with no protectable legal interests of their own.[18]

In opposition to those proposals, the following elements of a more life-protective legal framework for regulating the cryogenic preservation of human embryos is here proposed:

1. As already established in thirty-eight states, the life of the human being is expressly recognized to begin at fertilization or conception.
2. As already established in Louisiana, human embryos are deemed "juridical persons" entitled to pursue their continued development in the organism of their biological or adoptive mothers until birth. If IVF patients "fail to express their identity" or renounce parental rights, their physicians are deemed to be the embryos' temporary guardians until an adoptive implantation can occur.
3. As already established in New Mexico, human embryos can only be "disposed of" through implantation, not intentional destruction or through destructive human embryo research.

18. See Kimberly E. Diamond, "Cryogenics, Frozen Embryos, and the Need for New Means of Regulation: Why the U.S. Is Frozen in its Current Approach," *N.Y. Int'l. L. Rev.* 11 (1998): 77.

4. Prospectively, the cryopreservation of a human embryo shall only be permitted as needed to preserve life until implantation can occur.

5. Before another embryo may be conceived, already-existing embryos must be allowed to develop in the organism of their biological or adopted mothers until birth.

6. As already established in Australia, existing cryopreserved human embryos may be thawed and allowed to deteriorate after five years only under two conditions: (a) the gamete providers do not earlier consent to an extension of this time or to implanting the embryos in the organism of the biological or adopted mother, and (b) the custodian of the embryos cannot find anyone willing to accept implantation and assume parental responsibility for them.

7. Agreements involving the destruction of a human embryo or the donation of a human embryo for biological or medical research not designed to evaluate, protect, or restore the health of that human embryo, shall be unenforceable and not entitled to government funding of any kind.

8. No human embryo can be intentionally submitted to any exploitation whatsoever.

9. Prior to any IVF procedure, the gamete providers must sign a written disclosure form and dispositional agreement that fully informs them of the law requiring the implantation of all human embryos so conceived. This form must explain the costs, risks, and probabilities for success involved in IVF procedures being used. It must also identify the time for deterioration of cryopreserved embryos should such preservation be necessary to maintain the viability of any embryos prior to implantation.

The foregoing proposal offers a moral and juridical framework for decisions, decrees, and ordinances regulating the *ex utero* creation and use of human embryos. The case before us vividly illustrates the need for such a framework. This proposal is inspired by the compelling expert testimony of Dr. Jerome Lejeune — presented before the trial court in *Davis v. Davis* — that the human embryo is a human being.[19] In the words of the actual court's opinion:

---

19. See Dr. Jerome Lejeune, *The Concentration Can: When Does Human Life Begin? An Eminent Geneticist Testifies* (Ignatius Press: San Francisco, 1992). Part 5 of this work contains a "Proposal of Law on the Health of the Human Person," similar to the one proposed above, which was presented to the French Parliament on 20 May 1990 with the support of Dr. Lejeune.

[C]ryogenically preserved embryos are human beings. . . . Human embryos are not property. Human life begins at conception. Mr. and Mrs. Davis have produced human beings, in vitro, to be known as their child or children. For domestic relations purposes, no public policy prevents the continuing development of the common law as it applies to . . . human beings existing as embryos, in vitro, in this domestic relation case. The common law doctrine of parens patriae ["the power of the sovereign to watch over the interests of those incapable of protecting themselves"] controls children, in vitro. It is to the manifest best interests of the child or children, in vitro, that they be available for implantation. It serves the best interests of the child or children, in vitro, for their mother . . . to be permitted the opportunity to bring them to term through implantation.[20]

20. For a more complete discussion of the trial court's decision in *Davis v. Davis,* see the helpful booklet, "The Custody Dispute over Seven Human Embryos: The Testimony of Professor Jerome Lejeune, M.D., Ph.D.," available from the Christian Legal Society, 4208 Evergreen Lane, Annandale, VA 22003-3264.

CHAPTER 10

Bioethical Decisions When Essential Scientific Information Is in Dispute: A Debate on Whether or Not the Birth Control Pill Causes Abortions

Introduction

Linda K. Bevington, M.A.
Guest Editor for the Debate

Does the birth control pill ("the Pill") sometimes cause abortions by acting to prevent implantation of an early embryo rather than to prevent conception? This question has sparked a growing controversy regarding whether use of the birth control pill is morally permissible. Debate over this matter — which likely will escalate as awareness of the issue increases — is primarily fueled by a lack of consensus regarding the mechanism by which the Pill acts. Many physicians and patients would almost certainly reject the use of any contraceptive if they were convinced that it causes abortions.

Physicians and researchers who have devoted themselves to examining whether such a link exists have drawn markedly different conclusions as to whether current scientific evidence suggests that the Pill is indeed abortifacient. Although the authors of the debate presented here disagree over interpretation of the evidence, they are all committed Christians who recognize the magnitude of this issue and its potential implications for Pill-prescribers, Pill-takers, and, of course, the unborn. As a result, they have separately engaged in serious study and evaluation of the available evidence and have prayerfully reflected upon their findings.

This chapter was born out of the recognition that persons committed to upholding the sanctity of life should be provided with the opportunity to examine the evidence and arguments behind both positions in order to facilitate their own prayerful reflection on this critical matter. The authors were therefore invited to produce a manuscript defending their position on this issue. Each author team was provided with the other team's manuscript so that they could tailor their points to the arguments being presented by those holding the opposing view. Shortly before publication the authors were again given the opportunity to make changes to their manuscript before the final version was published.

In addition to laying out the opposing sides of the birth control pill issue, this debate is designed to address a larger question — namely, how should we develop an ethical position on a life-or-death issue when the scientific data required to draw a definitive conclusion is controversial or not yet available? The ethics of using the Pill, when some believe that it may act as an abortifacient, is merely a case in point. Even if scientific data were to resolve this particular matter tomorrow, the debate would still be illuminating, as we will always lack the data to definitively resolve certain ethical controversies. The authors warrant our deep gratitude and appreciation for their willingness to make this vitally important debate accessible.

Using the Birth Control Pill
Is Ethically Unacceptable

Walter L. Larimore, M.D.
Randy Alcorn, M.A.

Both of the present authors have independently reviewed the medical litera-ture concerning the potential abortifacient effects of the birth control pill ("the Pill"). We (like most in the Christian tradition) would define an "abortifacient effect" as an effect that ends the life of a pre-born child after fertilization or conception. Because we both advocated using the Pill for years, we hoped to discover that it does not cause abortions. Unfortunately, the scientific data led us to the opposite conclusion.[1]

1. R. Alcorn, "Does the Birth Control Pill Cause Abortions?" 3rd ed. (Gresham, Ore.: Eternal Perspective Ministries, 1998); W. L. Larimore and J. B. Stanford, "Postfertilization Effects of Oral Contraceptives and Their Relation to Informed Consent," *Archives of Family Medicine* (February 2000) (in press); W. L. Larimore, "The Abortifacient Effect of the Birth Control Pill and the Principle of Double Effect," *Ethics and Medicine* (January 2000) (in press).

Editor's Note: The Center for Bioethics and Human Dignity recognizes that decisions about prescribing and/or using the Pill should not be based solely on whether such an agent is abortifacient but, rather, should also take into account such factors as its effects on a woman's health and well being, and one's understanding of God's intended relationship between sex and reproduction.

Hostile Endometrium

Because space is limited, we will look at just two lines of evidence which indicate the Pill sometimes causes early abortions. First, a large number of medical studies document that the uterine lining (endometrium), the "home" new human beings implant in, is dramatically changed by the Pill. Although this is not direct proof of early abortions, it is indirect proof of high order.

This evidence is so well accepted in the medical world that the Food and Drug Administration's approved product information for the Pill in the *Physicians' Desk Reference* says, "Although the primary mechanism of action is inhibition of ovulation, other alterations include changes in the cervical mucus which increase the difficulty of sperm entry into the uterus, and changes in the endometrium which reduce the likelihood of implantation" (emphasis added).[2] This latter mechanism means that pre-born children may fail to implant in the uterus, at least on occasion, because of the Pill's effects. An independent clinical pharmaceutical reference also contains this assertion.[3] (Those who say these FDA-approved assertions are false should, in our opinion, prevail upon the FDA to change their statements and not simply ask people to disregard them.)

Reproductive endocrinologists have demonstrated that Pill-induced changes cause the endometrium to appear to be "hostile" or "poorly receptive" to implantation, at least in patients who are infertile.[4] One group of researchers has written that these "changes have functional significance and provide evidence that reduced endometrial receptivity does indeed contribute to the contraceptive efficacy of [the Pill]."[5] In other words, secular researchers consistently point out the abortifacient effect of the Pill. To date, no published studies have refuted these findings.

Magnetic Resonance Imaging (MRI) reveals that the endometrial lining of Pill users is consistently thinner than that of nonusers[6] — up to 58 percent

2. *Physicians' Desk Reference* (Montvale, N.J.: Medical Economics, 1998).

3. E. K. Kastrup, ed., *Drug Facts and Comparisons,* annual ed. (St. Louis: Facts and Comparisons, 1997).

4. H. I. Abdalla, A. A. Brooks, M. R. Johnson, A. Kirkland, A. Thomas, and J. W. Studd, "Endometrial Thickness: A Predictor of Implantation in Ovum Recipients?" *Human Reprod.* 9 (1994): 363-65.

5. S. G. Somkuti, J. Sun, C. W. Yowell, M. A. Fritz, and B. A. Lessey, "The Effect of Oral Contraceptive Pills on Markers of Endometrial Receptivity," *Fertil. Steril.* 65 (1996): 484-88.

6. J. M. Bartoli, G. Moulin, L. Delannoy, C. Chagnaud, and M. Kasbarian, "The Normal Uterus on Magnetic Resonance Imaging and Variations Associated with the Hormonal State," *Surg. Radiol. Anat.* 13 (1991): 213-20; B. E. Demas, H. Hricak, and R. B. Jaffe, "Uterine MR Imaging: Effects of Hormonal Stimulation," *Radiology* 159 (1986): 123-26;

thinner.[7] Recent and fairly sophisticated ultrasound studies have all concluded that endometrial thickness is related to the "functional receptivity" of the endometrium in women — at least in women undergoing fertility treatments.[8] Other studies have shown that when the lining of the uterus becomes too thin, implantation of the pre-born child (called the blastocyst or pre-embryo at this stage) does not occur.[9] The loss of a pre-born child is obviously abortifacient.

The minimal endometrial thickness required to maintain a pregnancy ranges from 5 to 13 mm,[10] whereas the average endometrial thickness in

---

S. McCarthy, C. Tauber, and J. Gore, "Female Pelvic Anatomy: MR Assessment of Variations During the Menstrual Cycle and with Use of Oral Contraceptives," *Radiology* 160 (1986): 119-23.

7. H. K. Brown, B. S. Stoll, S. V. Nicosia, J. V. Fiorica, P. S. Hambley, L. P. Clarke, and M. L. Silbiger, "Uterine Junctional Zone: Correlation between Histologic Findings and MR Imaging," *Radiology* 179 (1991): 409-13.

8. Abdalla et al., "Endometrial Thickness"; R. P. Dickey, T. T. Olar, S. N. Taylor, D. N. Curole, and E. M. Matulich, "Relationship of Endometrial Thickness and Pattern to Fecundity in Ovulation Induction Cycles: Effect of Clomiphene Citrate Alone and with Human Menopausal Gonadotropin," *Fertil. Steril.* 59 (1993): 756-60; Y. Gonen, R. F. Casper, W. Jacobson, and J. Blankier, "Endometrial Thickness and Growth during Ovarian Stimulation: A Possible Predictor of Implantation in In-vitro Fertilization," *Fertil. Steril.* 52 (1989): 446-50; L. B. Schwartz, A. S. Chiu, M. Courtney, L. Krey, and C. Schmidt-Sarosi, "The Embryo versus Endometrium Controversy Revisited as It Relates to Predicting Pregnancy Outcome in In-vitro Fertilization-Embryo Transfer Cycles," *Hum. Reprod.* 12 (1997): 45-50; Z. Shoham et al., "Is It Possible to Run a Successful Ovulation Induction Program Based Solely on Ultrasound Monitoring: The Importance of Endometrial Measurements," *Fertil. Steril.* 56 (1991): 836-41; N. Noyes, H. C. Liu, K. Sultan, G. Schattman, and Z. Rosenwaks, "Endometrial Thickness Appears to be a Significant Factor in Embryo Implantation in In-vitro Fertilization," *Hum. Reprod.* 10 (1995): 919-22; J. A. Vera, B. Arguello, and C. A. Crisosto, "Predictive Value of Endometrial Pattern and Thickness in the Result of In vitro Fertilization and Embryo Transfer," *Rev. Chil. Obstet. Gynecol.* 60 (1995): 195-98; J. H. Check, K. Nowroozi, J. Choe, D. Lurie, and C. Dietterich, "The Effect of Endometrial Thickness and Echo Pattern on In vitro Fertilization Outcome in Donor Oocyte-Embryo Transfer Cycle," *Fertil. Steril.* 59 (1993): 72-75; J. B. Oliveira, R. L. Baruffi, A. L. Mauri, C. G. Petersen, M. C. Borges, and J. G. Franco Jr., "Endometrial Ultrasonography as a Predictor of Pregnancy in an In-vitro Fertilization Programme after Ovarian Stimulation and Gonadotrophin-releasing Hormone and Gonadotrophins," *Hum. Reprod.* 12 (1997): 2515-18; C. Bergh, T. Hillensjo, and L. Nilsson, "Sonographic Evaluation of the Endometrium in In-vitro Fertilization IVF Cycles. A Way to Predict Pregnancy?" *Act. Obstet. Gynecol. Scand.* 71 (1992): 624-28.

9. Abdalla et al., "Endometrial Thickness"; Dickey et al., "Relationship of Endometrial Thickness"; Gonen et al., "Endometrial Thickness and Growth"; Oliveira et al., "Endometrial Ultrasonography as a Predictor"; Bergh et al., "Sonographic Evaluation of the Endometrium."

10. The 5 mm figure is from A. Glissant, J. de Mouzon, and R. Frydman, "Ultra-

women on the Pill is only 1.1 mm.[11] These data lend credence to the FDA-approved statement that "changes in the endometrium . . . reduce the likelihood of implantation."[12] If these indications are wrong, they should be refuted with evidence, not dismissed with wishful thinking.

Some physicians have theorized that when ovulation occurs in Pill-takers, the subsequent hormone production "turns on" the endometrium, causing it to become receptive to implantation.[13] However, there is no direct evidence to support this theory, and there is at least some evidence against it. First of all, after a woman stops taking the Pill, it usually takes several cycles for her menstrual flow to increase to the volume of women who are not on the Pill.[14] This suggests to most objective researchers that, rather than recovering quickly, the endometrium is slow to recover from its Pill-induced thinning. Second, the one study that has looked at women who have ovulated on the Pill showed that after ovulation the endometrium is not receptive to implantation.[15] In determining whether ovulation had occurred, this study looked for serum progesterone levels that are lower than those viewed by many (but not all) current researchers as indicators of ovulation. Therefore, critics of this study say it is possible that the 15 subjects studied with progesterone levels above 4 ng/ml simply had not ovulated. However, it is equally — or perhaps more — plausible to say: It is possible that all (or most, or even some) of the 15 subjects *had* ovulated." The bottom line is that this study, the only one of which we are aware, seems to refute the theory (unproved and never published in a peer-reviewed medical publication) that when a woman ovulates on the Pill her endometrium will *always* and completely normalize prior to implantation.

---

sound Study of the Endometrium during In vitro Fertilization Cycles," *Fertil. Steril.* 44 (1985): 786-90. The 13 mm figure is from R. Rabinowitz, N. Laufer, A. Lewin, D. Navot, I. Bar, E. J. Margalioth, and J. J. Schenker, "The Value of Ultrasonographic Endometrial Measurement in the Prediction of Pregnancy Following In vitro Fertilization," *Fertil. Steril.* 45 (1986): 824-28

11. McCarthy et al., "Female Pelvic Anatomy."

12. *Physicians' Desk Reference;* Kastrup, *Drug Facts.*

13. J. L. DeCook et al., "Hormone Contraceptives: Controversies and Clarifications," April 1999, available from ProLife Obstetricians, P.O. Box 81, Fennville, MI 49408, or by e-mail from prolifeob@aol.com.

14. J. B. Stanford and K. D. Daly, "Menstrual and Mucus Cycle Characteristics in Women Discontinuing Oral Contraceptives" (abstract), *Paediatr. Perinat. Epidemiol.* 9, no. 4 (1995): A9.

15. V. Chowdhury, U. M. Joshi, K. Gopalkrishna, S. Betrabet, S. Mehta, and B. N. Saxena, "'Escape' Ovulation in Women Due to the Missing of Low Dose Combination Oral Contraceptive Pills," *Contraception* 22, no. 3 (1980): 241-47.

Intrauterine/Extrauterine Pregnancy Ratio

A second line of evidence of the Pill's abortifacient effect is this: If the Pill has no post-fertilization effect, then reductions in the rate of intrauterine pregnancies in Pill-takers should be identical to the reduction in the rate of extrauterine (ectopic/tubal) pregnancies in Pill-takers. Therefore, an increased extrauterine/intrauterine pregnancy ratio would constitute evidence for an abortifacient effect.

Two medical studies allow review of this association.[16] Conducted at seven maternity hospitals in Paris, France,[17] and three in Sweden,[18] the studies evaluated 484 women with ectopic pregnancies and control groups of 389 women with normal pregnancies who were admitted to the hospital for delivery during the same time period. These studies were designed, in typical fashion for "case control" studies, to determine the risk factors for a particular condition (here ectopic pregnancy) by comparing one group of individuals known to have the condition with another group of individuals not having the condition. Both of these studies showed an increase in the extrauterine/intrauterine pregnancy ratio for women taking the Pill. Researchers who have reviewed these studies have therefore suggested that "some protection against intrauterine pregnancy is provided via the Pill's post-fertilization abortifacient effect," i.e., the loss of the pre-born child.[19]

That some other studies using a control group of non-pregnant women have not found a risk of increased ectopic pregnancy for Pill-users is not surprising. This apparent contradiction can be explained by the fact that contraceptive methods do indeed reduce the likelihood of pregnancy — whether normal or ectopic. Therefore, in determining whether there is an association between use of the Pill and an increased risk of ectopic pregnancy in women who get pregnant while taking the Pill, it is important that the subjects stud-

---

16. J. Thorburn, C. Berntsson, M. Philipson, and B. Lindbolm, "Background Factors of Ectopic Pregnancy. Frequency Distribution in a Case-control Study," *Eur. J. Obstet. Gynecol. Reprod. Biol.* 23 (1986): 321-31 (the original data was reevaluated by B. W. J. Mol, W. M. Ankum, P. M. M. Bossuyt, and F. Van der Veen, "Contraception and the Risk of Ectopic Pregnancy: A Meta Analysis," *Contraception* 52 [1995]: 337-41); J. Coste, N. Job-Spira, H. Fernandez, E. Papiernik, and A. Spira, "Risk Factors for Ectopic Pregnancy: A Case-control Study in France, with Special Focus on Infectious Factors," *Am. J. Epidemiol.* 133 (1991): 839-49.

17. Coste et al., "Risk Factors for Ectopic Pregnancy."

18. Thorburn et al., "Background Factors of Ectopic Pregnancy."

19. Larimore and Stanford, "Postfertilization Effects"; Thorburn et al., "Background Factors" (the original data was reevaluated by Mol et al., "Contraception and the Risk").

ied already be pregnant.[20] In our opinion (and in the opinion of the majority of the published studies done by secular researchers), the criticism of those who contend that the studies using pregnant controls are invalid is scientifically unfounded and ill-informed.

In other publications and in a much more detailed fashion, we have discussed these and other lines of evidence, citing dozens of scientific studies, as well as researchers and experts in numerous fields. We encourage interested readers to look more deeply into these studies and arguments.[21] Despite this evidence, some pro-lifers state that the likelihood of the Pill having an abortifacient effect is "infinitesimally low, or nonexistent."[22] Though we would both very much like to believe this, the scientific evidence does not permit us to do so. In our opinion, it would prevent most objective observers from doing so also.

Ethical Implications

When Christian physicians fail to reach a consensus, how should the Christian public respond? How should Christians make a decision — and how should pastors counsel church members — about taking the Pill, if the risk to unborn children is arguably real, but is believed by some not even to exist?

This is a significant question, since the Pill is used by about fourteen million American women each year and sixty million women internationally. Thus, even an "infinitesimally low" portion (say one-hundredth of one percent) of 780 million pill cycles per year globally could represent tens of thousands of unborn children lost to this form of chemical abortion annually. How many lives have to be jeopardized for pro-life believers to question the ethics of using the Pill? This is a question with profound moral implications for innumerable Christians.

The following are some of the common reactions we have encountered to the evidence that the Pill causes abortions.[23]

"I don't trust the evidence."

Most of those unsatisfied with the evidence do not appear to have closely examined it. Our natural tendency is to believe whatever supports a position in

20. Mol et al., "Contraception and the Risk"; Coste et al., "Risk Factors."
21. Alcorn, "Does the Birth Control Pill Cause Abortions?"
• 22. J. L. DeCook and J. McIlhaney et al., *Hormonal Contraceptives: Are They Abortifacients?* (Sparta, Mich.: Frontlines Publishing, 1998).
23. Alcorn, "Does the Birth Control Pill Cause Abortions?"

which we have vested interests and to disbelieve whatever contradicts it. We should resist this tendency and allow the evidence to determine our position — no matter how uncomfortable it makes us. We must each ask, "If this evidence doesn't convince me, is there any evidence that would? Is it possible that my vested interests in this issue are blinding me to the evidence? If there wasn't something to lose by believing and acting upon the evidence, would I still reject it?"

Note that scientists without pro-life convictions virtually always acknowledge that the Pill prevents implantation on occasion. Since this conclusion does not threaten them, they consistently come to it from the existing data. The evidence concerning the Pill is disturbing to pro-life Christians, but accepting it will help us make more informed and ethical choices.

"If we don't know how often abortions happen, why shouldn't I take the Pill?"

Imagine a hunter uncertain whether a movement in the brush is caused by a deer or a child. Should the hunter shoot? Imagine a woman driving at night who sees a dark figure ahead on the road, which may be a child or just a shadow. Should she continue to drive or put on the brakes? What if there is a 50 percent chance that it is a child? 30 percent? 10 percent? 1 percent? How certain does she have to be before acting? Should not the benefit of the doubt go to life?

We believe the Pill should not be used or recommended to others unless and until it is proven safe for all unborn children. To date, that proof does not exist. Furthermore, based upon the ethical principle of double effect,[24] one should not consider as ethical an action that *may* cause harm to human life (a bad effect) if there is another way of producing the good effect.[25] In the case of elective birth control, effective but non-abortificient contraceptive methods are available, and these include more than just barrier methods. Multiple studies over the past twenty years have proven the Creighton University approach to natural family planning to be even more effective than the Pill.[26] In a more recent controlled study, the researchers demonstrated that (per 100 couples over

24. W. T. Reich, "Double Effect," in Reich, ed., *Encyclopedia of Bioethics*, rev. ed. (New York: Simon & Schuster, 1995), 2:636-41.

25. J. Keown, "'Double Effect' and Palliative Care: A Legal and Ethical Outline," *Ethics and Medicine* 15, no. 2 (1999): 53-54.

26. T. W. Hilgers and J. B. Stanford, "Creighton Model NaProEducation Technology for Avoiding Pregnancy. Use Effectiveness: A Meta-Analysis," *J. Reprod. Med.* 43 (1998): 495-502.

one year) the pregnancy probability was 0.14 for perfect use with the Creighton method and 2.72 in actual use — an efficacy at least as great as the Pill.[27] In addition, the Billings Ovulation Method of natural family planning is so simple that it is taught around the world to people who cannot read or write.

Even if there were no effective alternatives to the Pill, it should not be used because the evidence suggests that it sometimes causes abortions. But the fact is, there *are* effective non-abortifacient alternatives to the Pill. Based upon the principle of double effect discussed above, it seems reasonable to conclude that the Pill should not be used or recommended to those who believe life begins at conception — unless and until the Pill is proven *not* to be an abortifacient. It appears that such studies could be done; however, to date, the needed proof clearly does not exist. Until such proof is available, the Pill should be considered a possible cause of death for pre-born children and therefore cannot be ethically used, recommended, or prescribed by Christians for contraceptive purposes.

"Spontaneous miscarriages are common — early abortions aren't that significant."

This argument boils down to: "Since God permits or causes millions of spontaneous abortions each year, it is acceptable for us to cause abortions too." There is a big difference, a cosmic difference, between God and us! What God is free to do and what we are free to do are not the same. God is the Creator; we are the created. As the maker of human life, God has the right to take it. We do not. Spontaneous abortions in women not taking contraceptives are not our responsibility. Abortions caused by chemicals we choose to take or prescribe are.

"The Pill's alteration of the endometrium hasn't been proven to cause abortions. It's only a theory."

No one who is aware of the facts can honestly argue that the Pill does not significantly change the endometrium. Such a change is undeniable. Accordingly, imagine a farmer who has two places where he might plant seed. One is rich, brown soil that has been tilled, fertilized, and watered. The other is hard, thin, dry, and rocky soil. If the farmer wants as much seed as possible to take hold and grow, where will he plant the seed? The answer is obvious.

27. M. P. Howard and J. B. Stanford, "Pregnancy Probabilities during Use of the Creighton Model Fertility Care System," *Archives of Family Medicine* 8, no. 5 (September-October 1999): 391-402.

Some physicians correctly point out that some newly conceived children do manage to implant and temporarily survive in hostile places, such as the fallopian tube. Elsewhere we explain this apparent contradiction.[28] But this in no way changes the obvious fact that many *more* children will implant and survive in a richer, thicker, more hospitable endometrium than in a thinner, less hospitable one.

"If this is true, why haven't I heard it?"

There are many answers to this question, including: (1) Concerns about early abortions are not widespread, since preventing implantation is not of any ethical concern except to those who believe human life begins at conception. (2) Published reports of Pill-caused abortions are available, but are spread out in dozens of obscure and technical scientific journals. (3) Medical semantics have played a role in obscuring the Pill's abortifacient mechanism. In 1976, the word "contraceptive" was redefined by the American College of Obstetricians and Gynecologists (ACOG) to include agents which prevent implantation. (4) The Pill is a multibillion dollar worldwide industry. Its manufacturers have tremendous vested interests in maintaining its status quo, as do most physicians, including pro-life Christians, who prescribe it.

"Using the Pill will mean fewer children die in spontaneous abortions."

Over one-half (some estimates range as high as 78 percent) of fertilized embryos are eventually spontaneously aborted.[29] Some pro-lifers point out that the Pill, by lowering the rate of conception, will lower the total number of deaths due to spontaneous abortions. The logic seems to be that when we use a chemical that kills *some* children, we can take consolation in knowing that this same chemical may prevent many other children from ever being conceived and therefore from dying "naturally."

This is convoluted logic, and again it puts us in the place of God. If there are fewer miscarriages because of the Pill, it is not because the Pill brings any benefit to a pre-born child, but only because it results in fewer children being conceived. It is not that lives are being preserved, but simply that there are

---

28. Larimore and Stanford, "Postfertilization Effects of Oral Contraceptives."

29. American College of Obstetrics and Gynecology, "Preembryo Research: History, Scientific Background, and Ethical Considerations," ACOG Committee Opinion 136. (Washington, D.C.: ACOG, 1994).

fewer lives to preserve. There is less death only because there is less life. Again, we must take responsibility for our choices, not for God's.

"Without the Pill there would be more elective abortions."

Suppose for a moment this were true. What is the logic? "If we go ahead and take action that will kill some children now, at least there may be other children, more of them, who won't get killed." The same approach could be used to deny drowning children access to a crowded life raft. This sort of pragmatism rings hollow when we put certain human lives at risk, without their consent, for the supposed good of others. However, the premise here is not true, as the small minority who would reject the Pill because it causes early abortions are the same people who would be the last ones to get a later abortion.

"Pill-takers don't intend to have abortions."

This is undoubtedly the case. Likewise, most Pill prescribers do not intend to cause abortions. Nevertheless, while the intentions of those taking or prescribing the Pill may be harmless, the results can be just as fatal. In this sense, taking the Pill is analogous to playing Russian roulette, but with more chambers and therefore less risk per episode. In Russian roulette, participants usually do not intend to shoot themselves. Their intention is irrelevant, however, because if they play the game long enough they cannot beat the odds — eventually someone dies. However, with Pill roulette, it is another person who may die. The fact that a woman will not know when a child has been aborted in no way changes whether or not it happens. The longer she takes the Pill, the greater her chance of having a silent abortion. The more a physician prescribes the Pill, the more likely he or she is to cause and be morally responsible for unrecognized abortions.

"I'm still comfortable prescribing the Pill because it's primarily contraceptive and only secondarily abortive."

Even if the Pill does not usually cause an abortion, whenever it does it is just as real an abortion as if that were its primary effect.

"You can't avoid every risk."

Indeed, some risks are necessary. But Pill-takers unnecessarily put their pre-born children at risk. In fact, the very survival of these children is at stake. Regardless of the actual risk percentage, which is uncertain, a sexually active woman runs a new risk of aborting a child every time she takes the Pill. Furthermore, as discussed above, she has non-abortifacient options for birth control, such as modern natural family planning, that may be more effective than the Pill.

"There may be a risk of the Pill causing an abortion, but if so, it is very small."

No one knows how small it is, but suppose it *is* very small. How much risk is acceptable risk? The answer, as discussed above, depends on the alternatives. There is no such thing as a car or a vaccine that poses no risk to one's children. But there is such a thing as a contraceptive method that does not put a pre-born child's life at risk. There are effective alternatives, such as natural family planning, that are as effective as — or more effective than — the Pill that do not and cannot cause abortions. It makes ethical, moral, and practical sense for Christians to learn more about these non-abortifacient birth control options. Unfortunately, physicians who ignore or deny the evidence indicating the abortifacient potential of the Pill are not likely to themselves seek such alternatives or to encourage their patients to do so.

"We shouldn't lay guilt on people by talking about this."

Scripture makes it clear that we are capable of doing wrong even when we are not consciously aware of it (Lev. 5:14-18; Job 13:23; Pss. 19:12; 139:23-24). We will each give an account to God of all we have done (Rom. 14:10; 2 Cor. 5:10). By coming to terms now with our sin and responsibility, we can to a certain extent preserve ourselves from having to face later judgment (1 Cor. 11:31). Because of the work of Jesus, God freely offers us pardon for everything — including actions taken in ignorance and sincerity that may have terrible and unintended results (Ps. 103:10-14). But to experience that forgiveness, we need to confess, repent of, and turn from our sin (1 John 1:9).

Some assume that one cannot set forth the evidence about the Pill without being guilty of legalism or spiritual abuse. Those who offer this argument do not give either truth or people enough credit. Is truth devoid of grace? Christ was full of grace *and* truth — so should we be. Are Christians incapable of handling difficult information or accepting God's provision for guilt? Is it compassionate to hold back disturbing truth from people rather than share it with them so that they can make their own decisions about what to believe and seek the Lord's guidance as to how to respond? The Christian life is not based on avoiding the truth but on finding and submitting to it.

"We shouldn't tell people the Pill may cause abortions until we know for sure."

Informed consent is a widely accepted ethical mandate of modern medicine.[30]

The physician's failure to provide adequate information seriously jeopardizes the patient's ability to make an informed decision. If this information is consciously withheld, it is a breach of ethics. Not telling women about the Pill's potential abortifacient effect betrays a disrespect for their intelligence, their moral convictions, and their ability to weigh the evidence and make choices. Not teaching them about non-abortifacient options such as natural family planning is ethically indefensible.

If physicians and pastors make their patients and parishioners aware of evidence indicating the Pill may cause early abortions and later research indicates that this evidence was not valid, what will have been lost? Informed decisions will have been made based on all the available data, and women will have had an excellent option in natural family planning. But if physicians or pastors fail to disclose the evidence to those in their care and it turns out it was valid all along, then they will have withheld vital information that might have kept pre-born children from dying. If we really love our most vulnerable neighbors — the unborn children — we will want to protect and preserve them instead of imperil them through our silence.

Further study of this matter is indeed necessary. We would be delighted if such study contradicted the existing evidence and demonstrated that the Pill is incapable of causing even a single early abortion. However, unless and until such study surfaces, the evidence that the Pill causes unrecognized abortions — at least some of the time — is cumulatively so substantial that we dare not ignore it.

It is hard to imagine a more horrid irony than that followers of Christ would speak out against surgical abortions, yet repeatedly make choices that result in the chemical abortions of their own children. If the Pill causes abortions, then the biggest threat to Satan's success is that people become aware of this truth and act on it. The evil one's vested interests in our blindness on this issue cannot be overstated.

It is not easy for pastors to speak out on this issue in their churches or for physicians to discuss it with their patients. However, pastors and physicians often have to address unpopular subjects. Why not this one? Some people will be angry and defensive, but most will be thankful and appreciative.

30. T. L. Beauchamp, "Informed Consent," in *Medical Ethics,* ed. R. M. Veatch (Boston: Jones & Bartlett, 1989), pp. 173-200.

We owe people the truth, spoken in love (Eph. 4:15). "Speak up for those who cannot speak for themselves" (Prov. 31:8). The issue is not whether people will applaud our decision to address this subject; the issue is whether the audience of One desires us to do so. If God does, all other opinions are irrelevant.

To us, the final word on this debate has already been given: "This day I call heaven and earth as witnesses against you that I have set before you life and death, blessings and curses. Now choose life, so that you and your children may live" (Deut. 30:19).

Using Hormone Contraceptives Is a Decision Involving Science, Scripture, and Conscience

Susan A. Crockett, M.D.
Joseph L. DeCook, M.D.
Donna Harrison, M.D.
Camilla Hersh, M.D.

Hormone contraceptives include combined oral contraceptives (COCs), injectables (DepoProvera), progestin-only pills (mini-pill, POPs), and the implant (Norplant).

The State of the Science

The idea that hormone contraceptives may occasionally cause a very early miscarriage or "mini-abortion" comes directly from the Food and Drug Administration-approved labeling requirements, arrived at by that government agency in cooperation with the manufacturer's research literature.[1] The primary mechanism of action of hormone contraceptives is to prevent ovulation; if this mechanism fails, sperm transport to the egg (and thus fertilization) will likely be impeded due to the contraceptive's thickening of the cervical mucus. Hormone contraceptive literature also reflects the fact that such contraceptives produce a lining of the uterus (the endometrium) that is less vascular and less glan-

---

1. *Physicians' Desk Reference* (Montvale, N.J.: Medical Economics, 1999).

dular than that normally seen six days after ovulation, when the fertilized egg would implant. The implication is that *if* a woman ovulates, and *if* the sperm gets through the thickened cervical mucus and fertilizes the egg, the altered uterine lining will make implantation difficult — and sometimes impossible — thereby causing the loss of the fertilized egg. This would indeed be a pre-implantation abortion. It sounds logical, and the hormone contraceptive literature implies it may happen. (Hormone contraceptive literature is written for marketing purposes ["this contraception *will* prevent pregnancy"] and for legal protection ["you can't sue us if you miscarry — we warned you"], as well as for patient education.)

However, this supposed abortifacient action is purely theoretical. The hormone contraceptive literature mentioning the altered uterine lining (the so-called "hostile endometrium") is only accurate if a woman has not ovulated. If she has not ovulated, there is no chance for pregnancy anyway. If a woman "breaks through" the contraceptive action and does ovulate, a whole new hormone environment comes into play, which has seven days to prepare the lining for implantation.[2] (Even if a woman is not using hormone contraceptives, on the day before she ovulates the lining of her uterus is also unfavorable to implantation, requiring this seven-day transformation.)[3] A burst of hormones, called the follicle-stimulating hormone and the luteinizing hormone, stimulates ovulation. From the day before through the days following ovulation there is an outpouring of natural estrogen (ten to fifteen times greater than hormone contraceptive levels — increased from 25 pg/ml to 250-400 pg/ml) and of natural progesterone (twenty times greater than hormone contraceptive levels —increased from 0.5 ng/ml to 10 ng/ml).[4] Over this period of days, these hormones from the corpus lutem of the ovaries transform the lining so that it becomes receptive to implantation. This is the physiologic effect of these hormone levels on endometrial tissue. In the event of ovulation, these hormones will be present transforming the uterine lining whether a woman is using hormone contraception or not (since hormone contraceptives are not known to suppress corpus luteum hormone output). This is likely the reason unexpected pregnancies in women using hormone contraceptives do as well as any other pregnancies.

The abortifacient theory is *not* a fact. It fails to account for the essential information about ovulation and its effect on the uterine lining. The concept of "hostile endometrium" is contrary to the known physiologic effect of ovulatory

2. Speroff, Glass, and Kase, *Clinical Gynecologic Endocrinology and Infertility,* 5th ed. (Baltimore: Williams & Wilkins, 1994).
3. Speroff et al., *Clinical Gynecologic Endocrinology.*
4. Speroff et al., *Clinical Gynecologic Endocrinology.*

estrogen and progesterone on the uterine lining. The FDA-approved labeling literature does not tell people this (the government's or manufacturer's literature does not always tell us everything we ought to know). Therefore, the misunderstanding is not the fault of concerned pro-life individuals who are misinformed by the inadequate, and even misleading, FDA-approved labeling literature.

We are aware of the study conducted in the mid-1970s at the Institute for Research and Reproduction of Bombay. This study claims to have demonstrated that fifteen women who ovulated while using hormone contraceptives each showed an endometrium that was atrophic — that is, less favorable to implantation by an embryo.[5] The researchers' criteria for proving that ovulation had in fact occurred was a progesterone level greater than 4 ng/ml. However, most medical experts agree that progesterone levels must be greater than 9 ng/ml to indicate the occurrence of ovulation.[6] This is a glaring weakness in the Bombay study and renders the endometrial biopsy findings meaningless, since it is very possible that the fifteen subjects studied with progesterone levels above 4 ng/ml simply had not ovulated.

The abortifacient theory proponents point out another line of evidence that they think suggests that hormone contraceptives are associated with an early abortifacient effect. They refer to an increased risk, per pregnancy, of ectopic (tubal) pregnancy.[7] The implication is that if there is an increased tubal pregnancy rate, then there must also be an increased number of embryos that enter the uterine cavity and are flushed out due to an inhospitable lining. The scientific literature does indicate an increased tubal pregnancy rate with progestin-only pills and Norplant. However, this increase is probably due to the progestin effect of slowing tubal motility — thereby increasing the likelihood of an embryo implanting in the fallopian tube rather than the uterus.[8] On the other hand, the literature on combined oral contraceptives and DepoProvera does not show an increased tubal pregnancy rate over normal.[9]

5. V. Chowdhury, U. M. Joshi, K. Gopalkrishna, S. Betrabet, S. Mehta, B. N. Saxena, "'Escape' Ovulation in Women due to the Missing of Low Dose Combination Oral Contraceptive Pills," *Contraception* 22 (1980): 241-47.

6. Shoupe et al., *Obstetrics and Gynecology* 73 (1989): 88-92; M. G. R. Hull et al., *Fertility and Sterility* 37 (1982): pp. 355-60.

7. W. L. Larimore, "Is the Birth Control Pill an Abortifacient?" WELS Lutherans for Life Bioethics Conference, Orlando, Florida, February 18, 1999, slides 49-61.

8. M. F. McCann and L. S. Potter, "Progestin-only Oral Contraception: A Comprehensive Review." *Contraception* 50 (1994): 21-53.

9. H. J. Tatum et al., "Contraceptive and Sterilization Practices and Extrauterine Pregnancy: A Realistic Perspective." *Fertility and Sterility* 28 (1977): 407-21.

Most abortifacient theory proponents have lumped together all of the hormone contraceptive agents — COCs, POPs, Norplant, and DepoProvera — as one type of agent. Problems related to rates of ovulation, ongoing pregnancy, ectopic pregnancy, and abortifacient action have been attributed to all four types alike. The result is a set of erroneous conclusions that impact ethical decision making regarding these medications. It is more instructive and accurate to review the literature concerning these agents separately since they vary in action, complications, and effectiveness.

The present authors have conducted a careful, exhaustive review of the medical literature regarding hormonal contraceptives' mechanism of action, considering the four types of hormonal contraceptives separately.[10] The discussion here summarizes their key findings.

We do not find substantive evidence of abortifacient action with use of hormone contraceptives. POPs are much less effective birth control than the other three types, although they have potential advantages for select patients.[11] POPs and Norplant are associated with higher ectopic pregnancy rates,[12] exposing the user to increased potential for morbidity and even mortality. This may constitute an unacceptable risk for the use of these products. On the other hand, the reviewed literature indicates that COCs and DepoProvera, with consistent and compliant use, are extremely reliable contraceptives. Their effectiveness depends on a high degree of ovulation suppression.[13] Secondarily, they alter cervical mucus to largely impede sperm penetration.[14] If these two mechanisms of contraception fail and conception occurs, the post-ovulatory hormone release from the corpus luteum would be expected to make the endometrium suitable for implantation. We find no evidence of abortifacient action and no demonstrable increase in ectopic pregnancy rates with COCs and DepoProvera.

The question of the mechanism of action of hormone contraceptives is a scientific one, which we as pro-life obstetricians-gynecologists have at-

10. The review paper is entitled "Hormone Contraceptives: Controversies and Clarifications." Copies of the paper, including a complete set of references, may be obtained on request from ProLife Obstetrician, P.O. Box 81, Fennville, MI 49408, or by e-mail from *prolifeob@aol.com*.

11. McCann and Potter, "Progestin-only Oral Contraception."

12. McCann and Potter, "Progestin-only Oral Contraception."

13. M. Toppozada et al., "Effect of Injectable Contraceptives Depo-provera and Norethisterone Oenonthate on Pituitary Gonadotropin Response to Luteinizing Hormone-releasing Hormone," *Fertility and Sterility* 30 (1978): 545-58.

14. F. C. Chretien, C. Sureau, and C. Neau, "Experimental Study of Cervical Blockage Induced by Continuous Low-dose Oral Progestogens," *Contraception* 22 (1980): 445-56.

tempted to address by a thorough review of the existing scientific literature. As discussed above, there is ample evidence for us to believe that certain forms of hormone contraception are not abortifacient.

In light of the above scientific discussion on hormone contraceptives and their mechanism of action, we will now take a moment and address some of the underlying issues that have prompted this research.

The controversy regarding the mechanism of action of the commonly used contraceptives has threatened to split the pro-life medical community. Review of information currently being disseminated reveals some powerful and well-written rhetoric. However, the question of hormone contraceptives' mechanism of action is not one which will be illuminated by rhetoric. The mechanism of action of any medicine will not change based on how we feel about the medicine, who developed it, or on how eloquently its use is defended or opposed. How a medication works is a scientific question, and at least for some of the hormone contraceptives, we have ample information to say clearly how and at what point in the reproductive cycle they exert their action. However, it is not so much the science as the theological and ethical questions that heat this debate within the pro-life community.

The Common Ground

The pro-life community begins this debate by agreeing that human life must be protected from its earliest beginning. As pro-life physicians in a culture which is aggressively pro-abortion, we have all felt the pain and grief associated with advocating respect for and protection of human life from its very beginning. We have all suffered the ad hominem attacks from pro-abortion colleagues, and we also have grieved as we watch our patients devalue themselves and destroy their children. The worldview of our pro-abortion colleagues is that human life is a product of random chance, as is animal life. So for them, our value as human beings comes from what we can do (especially what we can do for whoever is assigning us value). For the pro-abortion culture, our value is assigned to us by our human society and codified in law.

In sharp disagreement with this prevailing cultural worldview, Christians within the pro-life community typically look at human life differently. The value of human life is not dependent on the value of that human being to any other human being or human society. The value of any human life is measured by the value placed on that life by the One who created it. Therefore, from the beginning, the source of value for human life in the pro-abortion worldview and the pro-life Christian worldview are entirely different. Pro-life Christians typically value human life because God values human life in a way that is distinct from any other life that he created. And God's bestowed valuation of our life supersedes any assignment of value based on human choice, society, law, or any other human institution.

Christians agree that Scripture teaches that human beings are made in the image of God, by God, and for his purposes, and continue at his pleasure. We as human beings do not have the right before God to terminate the life of any other fellow human being, except as explicitly delegated by God in his Word. It is the love of Jesus Christ which constrains us to care for and love the human beings that he has made. God has designed that new human beings come into existence through, and in the context of, married sexual relationship. Marriage is a publicly witnessed, God-ordained act and institution of promise-giving and promise-keeping. At its core lies an accountability to God, humankind, and spouse for life-giving, truth-telling, faithfulness, and self-sacrifice. Within this refuge of safety lies the only biblical context for sexual intercourse and its designed outcomes, which include childbearing and childrearing. As are all of God's directives, this directive is for the highest good of the parents as well as of the children.

Human reason has been able to delineate the biological mechanisms by which God creates a new human being: the joining together of a male sperm and a female egg. The point in time at which a new human being is created is at the moment of conception. Therefore, any intentional interruption of the process of development after fertilization constitutes the moral equivalent of an abortion. Any intentionally caused abortion carries the same moral significance as the intentional taking of a human life at any time in the life span of that human being.

We are not free to use every means to achieve our goals. The things that we do must also not be contrary to God's revelation of his desires for our behavior. For example, it is good to have a child. However, if we commit adultery in order to achieve this good end, then God can and will judge us on the transgression of his moral law, regardless of our good intent. This truth has concrete implications for contraception and abortion. There are times when a conjugal union may occur without its primary purpose being childbirth. However, aborting any child simply because the child is unwanted is outside the limits of moral behavior that God has set, regardless of the "good" that is intended. Because Jesus' incarnation demonstrated that conception is the time at which God creates a human being in his image, we are constrained by the love of Christ to protect people from their conception to the end of their life as determined by God.

The Controversy

As demonstrated above, most pro-life Christians have much ground in common. However, within the pro-life community there are some who would contend that it is unethical to use any contraceptive mechanism whatsoever, while there are others who would allow for some righteous reasons to contracept. This essay is not meant to address this question directly. Rather, it necessarily makes the assumption that there may be some righteous reasons to contracept.

If there are righteous reasons to contracept, then are there righteous *means* to contracept? That is, are there contraceptives that work exclusively by preventing conception from occurring? And, given that biological systems are not perfect, do the contraceptives cause no harm to the child if a conception does occur? These are questions that the Christian pro-life community is compelled to face, both out of love for our Lord and out of love for his creatures. This is where the present controversy begins regarding the use of hormone contraception. Some sincerely motivated pro-life physicians have rejected the use of all hormone contraception as abortifacient, based on the following ethical principle: If even one abortion ever occurs in the whole history of the use of hormone contraception, then these contraceptives should never be used and the prescribers and the users are guilty of abortion.

The weakness of this principle is threefold. First, the structure of the statement makes it impossible to scientifically refute. Science is not designed to answer the question of whether or not, under any circumstances, a thing ever happens. Science is a method of modeling or describing reality based on serial observation. However, we cannot sample all of time or history. We cannot prove that an event never happened or never will happen. Yet that is the litmus test that our colleagues have proposed in order to accept the use of hormone contraceptives as a righteous means of contraception.

Second, let us examine this principle in the light of real medical decision making. Every medication that physicians prescribe, and every surgical procedure that they perform, carries with it real risks of injury or death. Yet, to withhold the medication or not to perform the surgery also carries real risks. If we try to make a decision based on our colleagues' proposed ethical principle, we will be paralyzed. For example, if a physician knows that one child in a million will die from administering a vaccine, then he or she should never vaccinate, because both the vaccinator and the parents who allow the vaccination would be guilty of murder. However, without the vaccine, the physician knows that 1 out of 100 children will die. So, then, the death of 1 out of 100 children is also the result of the physician's decision — the decision not to vaccinate. Although this is the greater evil, it may be easier for physicians to blame it on God's sovereignty. Somehow, to the best of their ability, the physician and the patient must make a judgment as to the balance between the known real risks versus the known real benefits. This vaccination example is useful to help clarify the medical decision-making process that accompanies any therapeutic action.

As illustrated above, any action carries with it real risks and benefits. Likewise, any inaction also carries with it real risks and benefits. How do a physician and patient decide on the best course of action for a patient? The decision making is an individualized process, based on all the information available to both the physician and the patient, to determine whether or not the benefits outweigh the risks. In this process, we ask the following questions:

- What are the known benefits of using a certain medication or procedure, and how frequently do those benefits occur?
- What are the known risks of using a certain medication or procedure, and how frequently do those occur?
- What are the known risks of *not* using a certain medication or procedure, and how frequently do those occur?
- What are the known benefits of *not* using a certain medication or procedure, and how frequently do those benefits occur?

Let us return now to the issue of hormone contraception and ask the four basic questions. (1) What is the benefit of hormone contraception? It is the ability to have sexual intercourse without the likely possibility of conception. (2) What are the risks of hormone contraception? They include health risks to the woman, the possibility of conception anyway, and health risks to the embryo/fetus. (3) What are the risks of not using hormone contraception? The most significant risk is an increased likelihood of conception if less effective or no alternative methods of contraception are used. This may result in an increased number of abortions, depending on the patient population. (Although our pro-life colleagues who maintain that hormone contraceptives are abortifacient have stated that abortions prevented by the use of contraception are of no consequence in this discussion, we disagree. Perhaps if a physician's patients are mainly Christians of impeccable moral life, then they would accept an unintended pregnancy as the merciful and gracious hand of a loving Father — which it is. However, in large populations of nominal believers or non-believers, many patients do not hesitate to say that if they found they were pregnant, they would abort.) (4) What are the benefits of not using a hormone contraceptive? The most important is the lack of health risks involved in using such a contraceptive. The present controversy revolves around the question of what are those health risks to the embryo/fetus.

Those opposing hormone contraception on the basis of the theoretical possibility of it being an abortifacient take the position that a theoretical, yet unproven, risk should carry the same weight and be disclosed in the same fashion as a known, proven risk. It is difficult to see where the line for such thinking will stop, because it depends on *who* thinks that a risk is theoretically possible. Proven risks are those which have recurringly happened in reality; for them, we can give a frequency of occurrence based on solid evidence from multiple studies over multiple periods of time. We can give the patient facts. Theoretical risks serve the very important function of defining future research directions, but we would be hard-pressed to compel colleagues to disclose risks that they do not think are real, and for which no real evidence exists to substantiate their occurrence.

Third, consider the implications if we generalize this principle to life. What if we were morally forbidden to take any action unless it could be proven under any and every circumstance that no harm will come to anyone at any time from that action? What if we have good reason to believe that doing a certain thing will be of benefit to someone else and have good reason to believe that it will not result in any harm, yet a bad consequence does occur? Is it better for us to abstain from doing good in fear that some evil may result — just so that the blame for such evil does not fall on us? Such fearfulness forgets the heart and core of our belief in the atoning sacrifice of Christ for our sins. This is not to suggest that we have a license to do bad things. Rather, it is to affirm the freedom which comes from the knowledge that our imperfect sinful attempts to do good with all our heart rest in final judgment at the feet of him who offers to atone for and redeem all things. And it is he who judges heart and mind, not according to appearances, but according to what is right. He is that Creator to whom we are accountable for our medical treatment of his creatures.

We recognize that equally spiritual and honest men and women may consider the same scientific data and hold the same ethical standards but come to differing conclusions on this matter. Thus we face, in the terminology of Romans 14, a "disputable matter" among believers. The principles of how to approach such a vital issue are laid out for us in Romans 14. We call attention to these principles without implication that one side or the other in this "disputable matter" is the "weaker brother," since we are all weak in many ways. Rather, we seek behavior and attitudes amidst this controversy that will "lead to peace and mutual edification" (v. 19).

First, we are commanded to accept one another "without passing judgment on disputable matters" (v. 1). We are instructed to "not look down upon" or "condemn" one another (v. 3), nor to "judge your brother" (v. 10). "For who are you to judge someone else's servant? To his own master he stands or falls. And he will stand, for the Lord is able to make him stand" (v. 4). Rather than judge others, we are instructed to be "fully convinced in our own minds" (v. 5) and to make decisions and perform actions "to the Lord" (vv. 6, 8). For "each of us will give an account of himself to God" (v. 12). It is noteworthy that either of the disputable behaviors discussed in these verses — eating or abstaining — may please or displease God, since God has other priorities: "For the Kingdom of God is not a matter of eating and drinking, but of righteousness, peace, and joy in the Holy Spirit" (v. 17). And finally, in our decision making on this issue, we must act in faith, for "everything that does not come from faith is sin" (v. 23).

The currently available technology is not sufficient to allow final and definitive scientific resolution of this controversy. However, abundant data are available for evaluation. For each individual, the decision becomes a matter of prayer, of evaluating sufficient pertinent scientific information, and of sensitivity to the Holy Spirit in decision making. As prescribing physicians, we have a special responsibility in this regard, since many of our patients, especially those with pro-life beliefs, will depend upon our evaluation of this vital subject in making decisions regarding their contraceptive choices. We are obligated to discuss with our patients sufficient factual information to enable them to give their "informed consent." If a couple decides to use hormone contraceptives as their method of family planning, it is our counsel that they are not using an abortifacient agent.

So, how are we as a pro-life community to respond to potentially divisive questions regarding the mechanisms of hormone contraception in the face of limited scientific evidence? We would draw a parallel to Paul's treatment of the issue of meat sacrificed to idols in 1 Corinthians 8–10. "So, whether you eat or drink, or whatever you do, do it all for the glory of God."

PART IV

RESPONDING TO
THE SEXUAL REVOLUTION

CHAPTER 11

Casualties of the Sexual Revolution: Youth Risk Takers

W. David Hager, M.D.

One important reason for today's widespread infertility, the desperate rush to reproductive technology, and a host of other social and health problems, is the risky sexual behavior of youth. Arguably the best way to prevent these problems is to address this behavior. Key steps in that direction are to appreciate the extent of youth risk taking today, to understand its causes and consequences, and to identify possible preventative measures.

Youth Risk Taking Today

As a recent Heritage Foundation study documents, from 1960 to the present we have seen amazing societal changes. In the United States, for example, there has been a 560 percent increase in violent crime. There has also been a marked increase in the out-of-marriage birth rate. We have seen a significant rise in the divorce rate, to nearly 50 percent. There has been a threefold increase in the number of children living in single-parent homes. The suicide rate among teenagers and children has increased by 200 percent. During the same period, SAT scores have dropped.[1]

At the same time, youth have increasingly engaged in behaviors that place them at risk for pregnancy outside of marriage, sexually transmitted diseases,

---

1. The Heritage Foundation, *Societal Changes in the United States, 1960-1995* (Washington, D.C. 1996).

and addiction to substances. The Youth Risk Behavior Surveillance Survey from the Centers for Disease Control and Prevention (CDC) details the health risk behavior of the youth of the United States.[2] It is a two-stage clustered sample taken from thirty-three states, three territories, and seventeen local school-based units. This 1997 survey reports the number of young people who:

- Never or rarely used seat belts within the last 30 days: 19.3 percent.
- Rode with a driver who had been drinking within the last 30 days: 36.6 percent.
- Drove a car after consuming alcohol within the last 30 days: 16.9 percent.
- Carried a weapon within the last 30 days: 18.3 percent.
- Carried a gun within the last 30 days: 5.9 percent.
- Were in a physical fight within the last 12 months: 36.6 percent.
- Were injured in a physical fight and needed medical treatment: 3.5 percent.
- Felt too unsafe to go to school on a daily basis: 4.0 percent.
- Carried a concealed weapon on school property while attending school within the last 30 days: 8.5 percent.
- Were threatened or injured with a weapon at school within the last 12 months: 7.4 percent.
- Had property stolen or damaged while going to school within the last 12 months: 32.9 percent.
- Considered suicide within the last 12 months: 20.5 percent.
- Attempted suicide within the last 12 months: 7.7 percent.
- Had ever smoked cigarettes: 70.2 percent.
- Smoked cigarettes within the last 30 days: 36.4 percent.
- Used smokeless tobacco within the last 30 days: 9.3 percent.
- Had ever used alcohol: 79.1 percent.
- Used alcohol within the last 30 days: 50.8 percent.
- Engaged in episodic heavy drinking within the last 30 days: 33.4 percent.
- Used marijuana within the last 30 days: 26.2 percent.
- Had ever used cocaine: 8.2 percent.

2. Centers for Disease Control and Prevention, "Youth Risk Behavior Surveillance United States, 1997," *Morbidity and Mortality Weekly Report* 47, no. 55-3 (1998): 1-89. See also Centers for Disease Control and Prevention, "Youth Risk Behavior Surveillance, National College Health Risk Behavior Survey — United States, 1995," *Morbidity and Mortality Weekly Report* 46, no. 55-6 (1997): 1-56. These reports also break the statistics down by gender.

The "stepping-stone theory" espouses that there is a progression of addictive processes which begins with nicotine and alcohol and progresses to harder drugs. Although this theory is debated, the fact that so many young people are smoking and drinking likely places them at significant risk for the use of drugs.

The data collected from adolescents who reported initiating risky behavior before thirteen years of age are also striking:

- Cigarette smoking, 24.8 percent
- Alcohol consumption, 31.1 percent
- Marijuana use, 9.7 percent
- Cocaine abuse, 1.1 percent

Sexual activity was also assessed in this survey. Among whites (non-Hispanic), 43.6 percent had previously had sexual intercourse and 32 percent were currently sexually active (within the last three months). Among blacks (non-Hispanic), 72.7 percent had previously had intercourse, and 53.6 percent were currently sexually active. Among Hispanics, 52.2 percent had engaged in intercourse, with 35.4 percent currently sexually active. Among students who had been sexually active, 27.8 percent said that they were now abstinent (within the last three months). Alcohol or drug use at their last act of intercourse was reported by 24.7 percent.

Data were further broken down according to grade in school:

Sexually Active	Grade 9	Grade 10	Grade 11	Grade 12
Ever	38.0%	42.5%	49.7%	60.9%
Currently	24.2%	29.2%	37.8%	46.0%

The number of young people who have had multiple sexual partners is also notable: 11.6 percent of whites, 38.5 percent of blacks, and 15.5 percent of Hispanics admitted having four or more sexual partners during their lifetime. In terms of grade level: 20.6 percent of twelfth graders, 16.7 percent of eleventh graders, 13.8 percent of tenth graders, and 12.2 percent of ninth graders admitted to four or more sexual partners during their lives. Having multiple sexual partners places the young person at a significant risk for pregnancy as well as all sexually transmitted infections and severe problems of emotional dysfunction.

Are the data from the CDC isolated? Do other studies indicate that teens are involved in sexual risk taking? Benson and Torpe have evaluated 976 Chicago eighth graders in terms of their history of having had sexual intercourse:[3]

History of Intercourse	Caucasians	Hispanics	Blacks
Males	42%	40%	82%
Females	13%	10%	32%

Benson and Torpe also evaluated a history of more than three sexual partners within the previous year.

More than 3 sexual partners within 1 year	Caucasians	Hispanics	Blacks
Males	41%	43%	64%
Females	11%	14%	22%

In another study, Small and Luster surveyed more than 2,100 young people in the seventh, ninth, and eleventh grades.[4] The study was not racially balanced, as only 8 percent were African Americans; the majority were Caucasian or Hispanic. The researchers graded the significance of risk factors. It is interesting that the authors found a statistically significant correlation among males and females for the top eight risk factors. The leading risk factors among both males and females were: (1) alcohol use, (2) steady girlfriend/boyfriend, (3) low parental monitoring, and (4) parents accepting adolescent sexual activity. Their data confirm the importance of parental monitoring and influence on young people.

The data also confirm earlier work by Donald Joy which indicates steady dating relationships among adolescents as being a predictor of early sexual debut.[5] This study correlated risk factors with sexual activity. The greater the number of factors present, the greater the chance that the young

3. M. D. Benson and E. J. Torpe, "Sexual Behavior in Junior High School Students," *Obstetrics and Gynecology* 85 (1995): 279-84.

4. S. A. Small and T. Luster, "Adolescent Sexual Activity: An Ecological Risk Factor Approach," *Journal of Marriage and Family* 56 (1994): 181-93.

5. Donald Joy, *Rebonding: Preventing and Restoring Damaged Relationships*, 2nd ed. (Napanee, Ind.: Evangel Publishers, 1996).

person was going to be sexually active. When there were no risk factors present, 15 percent of males and 1 percent of the females said that they were sexually active. Among females, if there were more than eight risk factors present, 80 percent were sexually active and among males if there were more than five risk factors present, 93 percent were sexually active. These data lend credence to the notion that we are not dealing with a single issue but a cumulative process. There are a number of issues that are interrelated, and altering risky behavior requires addressing them all.

Factors That Contribute to Risk Taking

What are some of the factors that may contribute to this explosion of risky behavior by young people? One of the greatest factors appears to be the media. The average teen watches twenty-three hours of television per week.[6] As Proverbs 23:7 says, "For as he thinks in his heart, so is he." If inappropriate information is fed into the brain, then inappropriate responses will result. If young people are exposed to sexually explicit types of behavior via the media, they are more likely to act out that behavior. The average teen watches seventy to ninety commercials per day. Many of the commercials on television today are sexually explicit or have profound sexual overtones. The advertisement of products seems to have degenerated to the level of associating every product with sexuality. The average teen sees 9,000 scenes of implied sexual intercourse per year on television with 80 percent of those outside of marriage. It is unusual for the consequences of sexual activity to be portrayed on television or in a movie. We seldom are told that the young woman who has engaged in sexual intercourse outside of marriage becomes infertile for the rest of her life because she has developed chlamydial salpingitis. We are not told that heterosexual activity is the most rapidly rising behavior by which HIV is acquired.[7] What we see is the pleasurable aspect of sexual interaction, without the potentially horrendous consequences.

In 1995, a *USA Today* survey reported that 83 percent of adults felt that the entertainment industry should decrease the incidence of sex and violence.[8] Sixty-eight percent said that the moral climate would improve if the industry did so, and 65 percent said that Hollywood is out of touch with the

---

6. "Teens and the Media," *USA Today,* 5 May 1995.
7. Centers for Disease Control and Prevention, "HIV/AIDS Surveillance Report — 1998," *Morbidity and Mortality Weekly Report* 47, no. 10-2 (1998).
8. "Teens and the Media."

moral values of the American people. Fifty-eight percent placed the blame on parents for allowing children to watch and participate in this type of activity. It appears the populace of this country feels that the entertainment industry is out of touch with what they want to teach their children. It is important that Americans transmit this information to the entertainment industry and emphasize their strong disagreement with the promotion of sexual activity outside of marriage.

Today teen magazines commonly feature articles with titles such as "Kissing 101 — Shy Kisses, Subtle Kisses, How Do You Kiss?" "Would You Rather Have Love or Sex?" and "Should Parents Let Their Kids Have Sex At Home?" This is what young people are faced with each day. They see this type of information, read it, and feel that it is normal to act it out. We have allowed the detestable to live with the good for so long that it has infiltrated our minds and we do not know the difference. We do not know what is bad and what is good. We commonly wonder, "Should I let my child engage in this behavior because it is so prevalent?" As discussed in other chapters of this book, we have to begin with a determination of what is morally right and wrong. What are God's moral standards? What does the Bible say about sexual activity outside of marriage? We must take a stand and be courageous in confronting these issues. The victims of this moral indifference are our children and our families.

Our family units are dissipated with a 50 percent divorce rate — many children living in single-parent homes, and many children without any parental influence or discipline. We are told that the government, not God, can supply the answers, can take care of all of these problems. Our young people are told that they can be protected from sexually transmitted diseases with a condom. Agencies such as SIECUS promote alternative sexual behaviors by indicating that one can avoid the consequences of engaging in these activities.[9] That sexually transmitted diseases can be transmitted by "outercourse" and that mutual masturbation frequently leads to more intimate sexual activity is ignored. We fail to teach our children delayed gratification. Instead, what they see is what they want and what they take. There has been a tremendous erosion of spiritual principles at school and in government. Unfortunately, the church has failed to address these issues adequately as well. It is easier to allow someone else to take care of the problem. It is not something we want to deal with ourselves.

---

9. SIECUS, *Guidelines for Comprehensive Sexuality and Education: Kindergarten–12th Grade* (New York, 1991), pp. 32-33.

Consequences of Risk Taking

The most visible consequences of non-marital sexual activity are sexually transmitted diseases and pregnancy. Of the ten leading infectious diseases in the United States today, five are sexually transmitted.[10] These STDs have a major impact on morbidity and health care expenditures. Young people comprise the largest subgroup of sexually active persons.

The CDC estimates that there are 12 million new cases of sexually transmitted infections in the United States each year. In the book, *Women at Risk: The Real Truth about Sexually Transmitted Diseases,* an annual figure of 15 million is cited.[11] Most of these cases — 10 million annually — occur among young people fifteen to twenty-nine years of age. A study conducted by the American Social Health Association Panel confirms this number.[12]

Various statistics are also available on the younger persons in this age range. Teenagers make up less than 10 percent of the United States population but they account for one-fourth of the sexually transmitted infections.[13] One in every four people who are newly infected with HIV is less than twenty-two years of age. Twenty-five percent of all young people twenty-five years of age or less have a viral sexually transmitted infection. Twenty percent of all sexually active teens get a new sexually transmitted disease (STD) every year. That means that there are 3 million teens newly infected on an annual basis. This is a major problem.

According to the CDC, adolescents ten to nineteen years of age and young adults twenty to twenty-four years of age are at the greatest risk for acquiring STDs.[14] There are several probable reasons: (1) They are more likely to have multiple partners. (To a young person, having one sexual partner commonly means, "I currently have one sexual partner." "One partner" can still mean that they are only serially monogamous.) (2) They are more likely to engage in unprotected intercourse. In spite of the numbers indicating that 50 percent of young men now use condoms, one-half are still not using condoms and frequently do not realize that condoms will not protect from many

10. Centers for Disease Control and Prevention. "Summary of Notifiable Diseases, United States — 1997," *Morbidity and Mortality Weekly Report* 47, no. 54 (1998).

11. David Hager and Donald Joy, *Women at Risk: The Real Truth about Sexually Transmitted Diseases* (Andersen, Ind.: Bristol House Ltd., 1994).

12. "American Social Health Association, Panel to Estimate STD Incidence, Prevalence and Cost," *Journal of the American Social Health Association* (1998): 6.

13. Centers for Disease Control and Prevention, "Infectious Diseases in the United States," *Morbidity and Mortality Weekly Report* 46 (1997): 257.

14. "Infectious Diseases in the United States," p. 257.

STDs. Even those who use condoms may not use them with every act of inter-course; nor are the eight steps of consistent and correct condom usage necessarily followed. (3) They may select partners at high risk. Most young people who engage in risky behaviors will select partners who engage in risky behaviors as well. (4) Teen women are especially susceptible physiologically to some organisms as long as the junction between the squamous and columnar types of cells that line the vagina and cervix/uterus is located on the outer portion of the cervix, where it is more susceptible to infection by bacteria like chlamydia.

Of the various reasons, the principal reason for the explosion of STDs among young persons is exposure to bacteria and viruses through multiple sexual partners. Human papilloma virus (HPV) is such an organism. HPV has become the most prevalent sexually transmitted virus in the United States today. According to a recent three-year survey evaluating HPV incidence among college coeds, the cumulative thirty-six-month incidence is 43 percent. Over the duration of the survey, 60 percent of the women were infected with HPV at one time or another. These young, intelligent college women were victimized by infection with HPV from males who probably had no idea that they were infected. These young women had engaged in risky sexual behavior with partners who were infected with a virus for which there is no definitive cure.[15]

The herpes family of viruses also poses serious problems.[16] The most familiar forms of this virus are the herpes simplex viruses (HSV) I and II which can cause genital herpes. Herpes virus 8 is an organism recently reported for its association with disease in gay males. A recent study found that 37.6 percent of homosexual males and none of heterosexual males were positive for this virus. This virus is now associated with the development of Kaposi's sarcoma in HIV-positive males. It is not just HIV that is lethal. The herpes virus enhances the chance of an HIV-infected male developing Kaposi's sarcoma and dying of the disease.

The problems and challenges of sexually transmitted diseases are enormous. Immunization, however, does not hold immediate promise. According

15. X. W. Sun, L. Kuhn, T. V. Ellerbrock et al., "Human Papilloma Virus Infection in Women Infected with the Human Immunodeficiency Virus," *New England Journal of Medicine* 337 (1997): 1343-49; G. Y. F. Ho, R. Bierman, L. Beardley et al., "Natural History of Cervicovaginal Papilloma Virus Infection in Young Women," *New England Journal of Medicine* 338 (1998): 423-27.

16. J. N. Martin, D. E. Ganem, D. H. Osmond et al., "Sexual Transmission and the Natural History of Human Herpes Virus 8 Infection," *New England Journal of Medicine* 338 (1998): 948-54.

to the Fourth Annual AIDS Conference held in Geneva, Switzerland, we are not very close to the release of a vaccine to immunize persons against HIV.[17] We are not any closer with regard to HPV or HSV. Moreover, a number of different non-viral organisms are developing resistance to treatment. *Trichomonas vaginalis,* a common cause of vaginitis in women, shows evidence of relative resistance to metronidazole.[18] There is even evidence that *Chlamydia trachomatis* may be developing resistance to tetracyclines. The key, then, is to prevent disease by limiting exposure. We cannot continue to develop antimicrobials that will effectively treat all organisms in all situations. We do not have the capability at the present of immunizing at-risk persons against viral sexually transmitted organisms. We must emphasize abstinence outside of marriage, and if individuals choose to ignore this sage advice, the use of condoms with every act of coitus. Along with this counsel, we must ensure that young people understand that condoms are not effective in preventing STDs transmitted by secretions, such as syphilis, HPV, and HSV.

Another consequence of risky behavior is non-marital pregnancy. There are over one million pregnancies among teens in the United States each year.[19] Ninety-five percent of these are unplanned. Fifty percent are pregnancies that occur within the first six months in which a teen has sexual intercourse. Forty-two percent of the time a pregnant teen mother chooses to abort the pregnancy. That number has been fairly steady for the last four to five years. Of teens who give birth, 83 percent end up on welfare. One-third of the fathers of these teen mothers are unemployed and not in school, and cannot provide for their own children. Moreover, there are much higher rates of divorce among this group than in the general population.

Several psychosocial factors contribute to young women becoming pregnant, including dependency needs, insecure feminine identity, lack of self-esteem, impulsiveness, attempting to replace an absent father, responding to stress, and having a pregnant sibling in the home.[20] If one teenager within the home becomes pregnant, and she has a sister who is close in age, there is a 75 percent chance that her sister will become pregnant as well. Data from Cal-

17. Centers for Disease Control and Prevention. *CDC Recommendations for Preventing Sexual Transmissions of HIV and Other STDs,* CDC update 2 (7 April 1997).

18. David Hager, "Trichomonas Vaginalis Ob-Gyn Infectious Disease Identification," in *Obstetric and Gynecologic Infectious Disease,* ed. J. Pastorek (Rockville, Md.: Raven Press, Ltd., 1994), pp. 537-43.

19. R. A. Maynard, ed., *Kids Having Kids: A Robin Hood Foundation Special Report on the Costs of Adolescent Childbearing* (New York: Robin Hood Foundation, 1996).

20. P. Donovan, "Falling Teen Pregnancy: What's Behind the Decline," *The Guttmacher Report on Public Policy* (New York: The Alan Guttmacher Institute, 1998).

ifornia analyzing adult paternity indicate that the average age difference between the sexual partner of a pregnant teen and the teen herself is 6.4 years.[21] Rates are similar across the country. There is a lack of parental discipline in the home to effectively prevent young girls from dating men much older than they are. The male is more mature physically, desires sexual gratification, and pressures the younger female to satisfy his sexual needs. Males in this situation usually abandon their teen girlfriends when they becomes pregnant. Characteristically, these males do not have a job and are not in school. That is why 83 percent of teen mothers end up on welfare.

There are serious consequences for the children of these teenage mothers as well. They are far more likely to be physically abused, neglected, or abandoned. They perform much worse in school: they are 50 percent more likely to repeat a grade, they perform significantly worse on tests of cognitive function, and they are far more likely to drop out of school.[22] The daughters of adolescent mothers are 83 percent more likely to become pregnant themselves. The sons of adolescent moms are 2.5 times more likely to serve time in prison. Thus we see the ripple effect of nonmarital teenage pregnancy. The cost to the United States alone in dealing with this problem is 29 billion dollars annually.

Prevention of Risk Taking

Jim Donovan has observed that sexual intercourse is an element of a syndrome of problem behaviors that includes drug, alcohol, and nicotine use as well as school difficulty plus minor delinquency. He has aptly termed this "a conundrum of problems."[23] A study in the journal *Pediatrics* evaluated several risky behaviors and found that the rates for each risk factor were significantly higher among non-virgins as opposed to virgins.[24] Young people who are engaged in one form of risky behavior are much more likely to also be sexually active. The United States has tried to discourage risky behavior by raising cigarette taxes, proposing drug emphasis programs, and attempting to create certain barriers to involvement in activities that predispose the young person to addictive behavior. The government has also developed legislation

21. Maynard, ed., *Kids Having Kids*.

22. Maynard, ed., *Kids Having Kids*.

23. J. E. Donovan and R. Jessor, "The Conundrum of Teen Sexuality," *American Journal of Public Health* 73 (1983): 543-52.

24. D. P. Orr, M. Beiter, and G. Ingersoll, "Premature Sexual Activity as an Indicator of Psychosocial Risk," *Pediatrics* 87 (1997): 141-47.

to decrease the chance of a young person carrying a weapon onto school property. There are other efforts being made to decrease violence. Unfortunately, the United States has not made as much progress with the problems of pornography, alcohol use, gambling, and sexual activity among teens. There is tremendous concern over the use of nicotine among teenagers, but much less emphasis is placed on activities such as pornography that might promote sexual activity. Could this be true because adults who indulge in unhealthy sexual behavior are unwilling to talk to young people in an emphatic way about its dangers? Fear of hypocrisy is a great deterrent to action. An adult may think, "I'm having an affair, I'm into pornography myself, so how can I advise my child against this type of activity?" Young people need role models among the adult population who are living consistent lives and are willing to take a stand on the issues.

Young people are risk takers. They think that they are invincible. Many adults felt the same way when they were younger. It is imperative that we help young people to understand the risks that they are taking — that is, the potential consequences of their actions. We must teach young people to protect themselves and avoid harming others. They must face the fact that they cannot even trust each other. Studies indicate that 59-92 percent of college males surveyed said that they would lie to a potential sexual partner about having an STD in order to achieve sexual favors.[25] Many of these males may already be infected and have no idea that they are transmitting disease. Why would young people buy into this concept of safe sex and assume that a condom is going to help protect them? They feel that they are invincible and are not going to contract an STD, or become pregnant, or father a child. Programs must be developed to convince those who are engaged in risky behaviors that the very possible consequences of those behaviors are dire and to be avoided. Risk takers need to fully appreciate how likely it is that they will eventually contract an incurable disease, become permanently sterile, face an unplanned pregnancy, or even die as a result of making the wrong choices.

Contributing to the frequency of multiple partners among sexually active young people is the intricate process of bonding in relationships. In his book *Intimate Behavior*, Desmond Morris describes the twelve steps to bonding.[26] From step 1 (one person notices the other of the opposite sex) through

25. W. M. Burdon, "Deception in Intimate Relationships: A Comparison of Heterosexuals and Homosexuals/Bisexuals," *Journal of Homosexuality* 32 (1996): 77-93; G. J. Fischer, "Deceptive, Verbally Coercive College Males: Attitudinal Predictors and Lies Told," *Archives of Sexual Behavior* 25 (1996): 527-33.

26. Desmond Morris, *Intimate Behavior* (New York: Random House, 1971).

step 3 (verbal communication occurs), to more public displays of bonding such as step 5 (arm to shoulder) and step 6 (arm to waist), there is a progression of intimacy which is normal and appropriate. At step 7 (face-to-face contact) the first level of sexual arousal occurs. By step 9 (reclining body) the couple is comfortable lying together. Steps 10, 11, and 12 (mouth to breast, hand to genital, and genital to genital, respectively) are acts of sexual intimacy intended to bond for a lifetime. The steps of bonding, according to Morris, should occur in sequence. If steps are missed, the result will be a weak bond that is likely to break. Because each step results in more sensation, there is a desire to progress from one step to the next. Donald Joy also comments on the intimacy sequences of bonding in his book *Bonding: Relationships in the Image of God.*[27] He indicates that the sensations of each step are heightened by brain chemicals and neurotransmitters which seal each step in the person's memory. It is important to allow each step to be sealed before moving on to the next one.

Joy says that if a relationship is lost or abandoned, all future relationships move prematurely, often instantly to the highest step achieved in the previous relationship. This would account for the frequency of sexual intercourse in subsequent relationships after a previous one ends. Since the steps of bonding and intimacy are sealed in the memory, there will be significant emotional turmoil when the conflict of repeated intimate bonding occurs. The key, then, is to prevent progressing through so many of the steps of intimacy outside of marriage, saving a number of those steps for a relationship of commitment for a lifetime.

How can the principles of abstinence be conveyed to young people who desire to take risks? Teaching from a knowledge-based perspective has not been successful. Using the principles of character-based sexuality education, such as those published by the Medical Institute for Sexual Health, is key.[28] Programs must be comprehensive to be effective — covering all of the risky behaviors including the use of nicotine, alcohol, and hard drugs, as well as pornography and sexual activity. We must emphasize that the steps of intimate sexual bonding are to be reserved for marriage. We must teach young women and men refusal skills so that they are prepared to say, "NO." Programs such as that developed by McCabe and Howard at Emory University prove that rates of intercourse and teen pregnancy can be reduced by doing

27. Donald Joy, *Bonding: Relationships in the Image of God* (Waco, Tex.: Word Books, 1985).

28. "National Guidelines for Sexuality and Character Education" (Austin, Tex.: Medical Institute for Sexual Health, 1996).

so.[29] It will take a dedicated effort on the part of educators and health care professionals, but it is an endeavor that is well worth the work.

We must also emphasize the role of parental involvement. Parental presence in the home, parental expectations, and parental disapproval of early sexual debut are crucial. We must encourage adults not to live lives of hypocrisy. Many adults tell their children to do one thing while living out a different lifestyle. Our youth must have appropriate role models and examples of Christian living in order to deal with the issues that they face. Parental direction is the greatest deterrent to delinquent sexual behavior, according to the ADD Health Study.[30] We must encourage parents to become involved in instructing their own children and not depending on others to do so.

We live in a critical time. A concerted and comprehensive effort to address youth risk taking is essential if we are to adequately protect our children and our future.

29. M. Howard and J. B. McCabe, "Helping Teenagers Postpone Sexual Involvement," *Family Planning Perspectives* 22 (1990): 21-26.

30. M. D. Resnick, P. S. Bearman, R. W. Blum et al. "Protecting Adolescents from Harm: Findings from the National Longitudinal Study on Adolescent Health," *Journal of the American Medical Association* 278 (1997): 823-32.

CHAPTER 12

Sex in America:
Past, Present, and Future

Joe S. McIlhaney Jr., M.D.

"America has a message about sex, and that message is none too subtle. Anyone who watches a movie, reads a magazine, or turns on a television has seen it. It says that almost everyone, but you, is having endless, fascinating, and varied sex." So observe University of Chicago researcher Robert T. Michael and his co-authors in their book, *Sex in America*.[1] How did we get to our present sex-saturated state? What have been the results? Where do we go from here?

This is no academic issue for me. I left a fascinating practice of gynecology and reproductive medicine and committed myself to searching out and communicating the answers to these questions. A philosopher who spoke at a program I participated in recently said that we can look at the past to understand the present. I agree with him. We must do that to find answers to our questions. So, how did we get here? What forces produced the sexual revolution of the 1960s and, in turn, the sexual environment of today?

Competing Philosophies

As we look at the flow of thought about sexuality down through the centuries, we can see that two rival philosophies have shaped our culture's attitudes. These philosophies have been vying for dominance, not only in cul-

1. R. T. Michael, J. H. Gagnon, E. O. Laumann, and G. Kolata, *Sex in America* (Boston: Little, Brown, and Company, 1994).

ture, but in individuals' lives from the beginning. Surprisingly, though, in spite of their struggles with each other, both have added some very positive elements to our understanding of sexual issues today. Both philosophies are far more complex than can be discussed or documented here. Nevertheless, the difference between their outlooks on sexuality is so pronounced that even a brief survey of some proponents of each will prove instructive.

The first philosophy has greatly influenced human sexual practice from before the dawn of written human history until recent times. This stream of thought has been fed largely by the major religions of the world. The message is that sex, unsullied by the interference of people, is a very special gift from God, and that it must be used within certain guidelines in order for us to enjoy its benefits. However, if we do not use it within those limits, it can cause much pain and heartache.

One source of this stream is an array of Judeo-Christian writings passed down through the centuries. For example, according to the book of Genesis, "For this cause a man shall leave his father and his mother, and shall cleave to his wife; and they shall become one flesh" (2:2). One thousand years before Christ we hear these words in the Song of Solomon: "I have come into my garden, my sister, my bride. I have gathered my myrrh with my spice" (5:1). This is joyful, exuberant sex within marriage. Even the Puritans — a people of great faith who lived primarily in the seventeenth century and are often parodied as opposed to sexual pleasure — are described by Gottlieb as follows: "Puritan writers in particular, praised the delights of married love. . . . marriage and love naturally went together."[2]

Other major religions of the world disagree with Judeo-Christian teaching at numerous points. However, these disagreements do not generally include core sexual issues — all agree that sex should be reserved for marriage.

Buddhism has five major precepts, one of which is sexual purity. The Dalai Lama is today's most prominent Buddhist, and he writes very clearly of marriage being the place for sex.[3]

Hinduism also holds five rules to be of highest value. One of these rules, to which most Hindus strictly adhere, is that they not engage in adultery, which also includes refraining from premarital sexual activity.[4]

---

2. B. Gottlieb, *The Family in the Western World: From the Black Death to the Industrial Age* (New York: Oxford, 1993), p. 99.

3. A. Powell, *Living Buddhism* (Berkeley: University of California Press, 1995), p. 24. Also, the Dalai Lama, *The Power of Compassion* (San Francisco: Thorson, 1995), p. 59.

4. D. Kinsley, *Hinduism, a Cultural Perspective* (Englewood Cliffs, N.J.: Prentice-Hall, 1993). Also, V. P. Kantikar and O. Cole, *Teach Yourself Hinduism* (Chicago: NTC Publishing Group, 1995), p. 84.

Most know of the Muslim laws against sex outside of marriage. Few Westerners, however, know that Muhammad encouraged enjoyment of sex by both husband and wife, and that in discussing cruelty, he once cited intercourse without foreplay as a form of cruelty to women. Most do not know that Islam is one of the few religions to include sex as one of the rewards in the afterlife.[5]

The influence of religion on sexual practice has been profound and consistent, and most of this influence has been protective to individuals in society. Those who have practiced sexual activity according to the rules of their faith have maintained a faithful monogamous marriage. They have had intercourse only with their wife or husband for life. This practice has provided near-absolute protection from sexually transmitted disease and non-marital pregnancy. There have been other benefits from this practice, primarily for the woman. For example, she has usually not been left to provide for and raise a child without the help of a mate.

Though not everyone in the past or present are adherents to religious belief, there has been enormous benefit to all from the influence of the many faiths. They have been the fountain from which our concepts of character have come — the value of the individual, respect, responsibility, self-control, delayed gratification. These and others are the universal core ethical values that allow diverse peoples to live together in harmony and which can cause sex to be healthier, both physically and emotionally.

Then there is a second stream of thought. We can pick up the threads of this second philosophy in the early nineteenth century, because it was during this time that it began to influence modern thought.[6]

In 1812, Thomas R. Malthus theorized that a population tends to increase at a faster rate than its means of subsistence. Charles Darwin in 1864 was impressed by Malthus's theory, and his theory of evolution was significantly influenced by it. One of Darwin's intellectual heirs (and first cousins), Francis Galton, extended Darwinist thinking in a way that ultimately gave birth to the eugenics movement. Human eugenics means the selective breeding of human beings. This movement became very strong during the first twenty-five years of the twentieth century. The eugenics movement represented a great divergence from the teachings of most religions, which emphasized that every individual has immense intrinsic value.

---

5. G. Brooks, *Nine Parts of Desire: The Hidden World of Islamic Women* (New York: Anchor Hardcover, 1995), p. 39.

6. D. Freeman, *Margaret Mead and Samoa: The Making and Unmaking of an Anthropological Myth* (Cambridge, Mass.: Harvard University Press, 1983).

It was into this fertile new soil that Margaret Sanger, founder of Planned Parenthood, planted her new ideas. She wrote articles about eugenics such as one titled "Birth Control: To Create a Race of Thoroughbreds."[7] Influenced by this and other philosophies, she founded the first U.S. birth control clinic in 1916. Before the opening of the clinic, Sanger, who involved herself sexually with various men and women, had written, "The marriage bed is the most degenerating influence in the social order."[8]

Out of this context of fermenting actions and ideas emerged Margaret Mead's 1928 publication, *Coming of Age in Samoa*. This book was published as scientific proof that not only was unrestrained sex not "bad," but it was the healthiest sex. She concluded, for example, that the exceptionally smooth sex adjustment among adult Samoans is preceded by a time of free lovemaking and promiscuity among adolescents before marriage.[9] The book received wide acceptance. Martin Orans and other anthropologists have shown that Mead, a twenty-three-year-old graduate student who spent a total of nine months in Samoa and had a very poor grasp of the language, published findings that were at best naive, and at worst dishonest. Orans, in his book *Not Even Wrong*, in speaking of anthropologists' wide acceptance of Mead's work, wrote: "Doubtless many of us did so because we wanted such findings to be correct." He goes on to say, "Because Mead's work was deemed to possess authority, it became a guide for proper childrearing and justification for sharing more sexual knowledge with children and a more permissive attitude toward childhood and adolescent sexual practice. Certainly all those who already held such views welcomed Mead's validation."[10]

Mead was at the peak of her acclaim in the 1950s, and Sanger lived until 1966. At about this same time, two monumental, history-altering events occurred. The first of these was the advent of television. Prior to television most culture was local culture. Music, dance, stories, and other forms of entertainment primarily originated from family and neighbors. Values communicated through these cultural forms were generally the values of the local community. In contrast, most of our cultural life is now defined by distant strangers, and television has been central to the transition from a local to a mass-mediated culture. The family should serve as a screen for the child from the world, selecting what it deems worthy of attention by the

7. R. Marshall and C. Donovan, *Blessed Are the Barren* (San Francisco: Ignatius, 1991), p. 9.

8. Ibid., p. 7.

9. Freeman, *Margaret Mead*, p. 92.

10. M. Orans, *Not Even Wrong: Margaret Mead, Derek Freeman, and the Samoans* (Novato, Calif.: Chandler and Sharp, 1996).

child. The family is not the screen in the lives of children who are given large daily doses of television.

The second history-altering event of the sixties era was the release by the FDA in 1960 of the first birth-control pill. The point here is not to discuss ethical concerns about oral contraceptives. Rather, the issue is that for society in general, the Pill allowed the possibility of Mead's unrestrained sexuality and Sanger's feminine freedom to become reality. In retrospect, it is as though a number of uncoordinated lights all became laser focused on the early 1960s — to set fire to a philosophy that had been smoldering in society for many years — giving birth to the sexual revolution. The sexual revolution was a cataclysmic change in course for Western culture. With its advent came the sudden overwhelming dominance of sex without rules which resulted in "if it feels good do it" actions.

Some champions of the sexual revolution have dismissed the wisdom available to us from centuries past. For example, when Kristine Gebbie was a high-level AIDS official under President Clinton, she said the following: "As long as the prudish are allowed to define premarital sex in terms of don'ts and diseases, we will continue to be a repressed, Victorian society that misrepresents information."[11] The problem with this statement is the implication that values that protected sex, handed down to us from the past, especially if they come to us through religion, are dangerous. Rhetoric of this type has damaged the healthy balance that needs to be achieved in our society by using what is of value from both philosophical streams.

The Need for Boundaries

There have indeed been helpful contributions to our understanding and discussion of sexuality from both of the streams discussed here. For example, we all have become much better at speaking about sexual issues, not only privately, but also publicly. More parents are teaching their children about sexuality. Some of the information concerning sexual issues covered in this book could not have been discussed openly forty years ago. But this openness about sexual issues needs to be connected with a strong societal message that sex must be practiced within boundaries that protect the individual and society. For some people, these boundaries may be provided by their faith; for others faith may not be involved, but guidelines are necessary nevertheless. Realistic,

11. T. Strode, "Stressing Abstinence 'Criminal,' Gebbie Says," *Light* (November-December 1993), p. 14.

fair rules increase our freedom and joy, not diminish it. Consider something as simple as a football game. Football played without rules would be chaos. What we have seen in this country for the last forty years has often been openness about sex, but without the help of guidelines.

A telling incident during the 1990s involved a group of boys in California who called themselves the Spur's Posse. They were competing with each other to see how many girls they could have intercourse with. One of the boys responded to the criticism of their group by saying, "They pass out condoms [and] teach sex education . . . but they do not teach us any rules."[12] Theodore Roosevelt once said, "To educate a man in mind and not in morals is to educate a menace to society." It seems we are seeing the problems that result from just teaching about sex but ignoring the rules. Enormous problems have resulted from this imbalance. These problems are hurting people, especially the youth. Youth, such as the young women who were manipulated into sexual activity by the Spur's Posse, were emotionally hurt. Actually, so were the boys.

Indeed, the issue of sexuality in America is fraught with problems. While several other chapters in this volume present a fuller picture, a few key points are worth noting here.

- At least 15 percent of infertile women cannot have children because of damage caused by a sexually transmitted disease (STD).[13] STDs are the most rapidly increasing cause of sterility. Some in vitro fertilization programs would probably not be economically viable if it were not for STDs.
- A recent study published in the *New England Journal of Medicine* found that 45.9 percent of all African Americans over the age of eleven are seropositive for genital herpes. In addition, it showed that there has been a 500 percent increase in genital herpes among white teens in the past ten years. Approximately 20 percent of Americans over the age of eleven are now infected with this disease.[14]
- A recent study at Rutgers University has shown that over a three-year period, 60 percent of sexually active coeds were infected with human papilloma virus (HPV). This disease is the cause of more than 93 per-

12. J. Gross, "Where 'Boys Will be Boys,' and Adults Are Bewildered," *New York Times,* 29 March 1993, p. A1.

13. T. R. Eng and W. T. Butler, eds., *The Hidden Epidemic: Confronting Sexually Transmitted Diseases* (Washington, D.C.: National Academy Press, 1997), p. 45.

14. D. T. Fleming, G. M. McQuillan, R. E. Johnson, A. F. Nahmias, S. O. Aral, F. K. Lee, and M. E. St. Louis, "Herpes Simplex Virus Type 2 in the United States, 1976-1994," *New England Journal of Medicine* 337, no. 16 (1997): 1105-11.

cent of all truly abnormal pap smears and cervical cancer in the United States.[15]

- Almost as many Americans have died of AIDS as Americans who died in World War II, approximately 400,000 people.[16]
- Twenty percent of sexually active teens get a new sexually transmitted disease every year.[17]

Another aspect of this epidemic is the impact of non-marital teen pregnancies on young mothers, their children, and society. In addition to the physical and emotional damage done to so many, the social consequences are staggering:[18]

- Seven out of ten young adolescent mothers drop out of high school.
- The long-term wage-earning power of adolescent fathers is greatly reduced.
- The teenage sons of adolescent mothers are 2.7 times more likely to spend time in prison than are the sons of mothers who delay childbearing until their early twenties.
- The teenage daughters of adolescent mothers are 50 percent more likely to bear children out of wedlock.

These are overwhelming problems, damaging hundreds of thousands. Many students of culture feel that these are among the greatest problems society faces today. If we recognize the medical dimension of these challenges, we can more readily find constructive ways forward.

Sexually transmitted diseases and non-marital teen pregnancies are destroying the health and happiness of millions of people in the United States and around the world. Because these are primarily medical problems, their prevalence and impact are measurable. We have to guess about the impact of some societal problems, such as media's impact on children, but not about this epidemic. We can know how severe it is by measuring it. We can also

15. G. Y. F. Ho, R. Bierman, L. Beardsley, C. J. Change, and R. D. Burke, "Natural History of Cervicovaginal Papillomavirus Infection in Young Women," *New England Journal of Medicine* 338, no. 7 (1998): 423-28.

16. Centers for Disease Control and Prevention, *HIV/AIDS Surveillance Report: U.S. HIV and AIDS Cases Reported through December 1997* 9, no. 2 (1997): 19.

17. Alan Guttmacher Institute, *Facts in Brief: Teen Sex and Pregnancy.* Internet. (4 June 1998). Available at *www.agi-usa.org/pubs/fb_teensex.html.*

18. R. A. Maynard, ed., *Kids Having Kids: A Robin Hood Foundation Special Report on the Costs of Adolescent Childbearing* (New York: Robin Hood Foundation, 1996).

measure attempts at solving these problems. By doing this, we can tell what works and what does not work. For instance, are the numbers of adolescents who are infected with STD or pregnant out of wedlock decreasing dramatically as the result of various interventions or not?

Medically trained people know how to work with measurable medical problems. That is what we do. And this area of involvement is our responsibility. Policymakers, educators, and the media need to be accountable to the message of medical data that shows whether society's efforts are successful or not. If programs are actually working by dramatically lowering STD and nonmarital pregnancy rates, we need to embrace them. If not, we need to eliminate them. Until we do this, the epidemics will continue.

Ways Forward

Many constructive responses are possible. We need to invest time and resources to understand the challenge before us. In particular, we must learn more about those who are being hurt the most — our teens. They are more susceptible to STDs than adults, and they are the group in our society most highly infected with many STDs. They need the assistance of adults — adults who will love them enough to be directive and tell them what they can do to avoid these problems totally. Young people want to know what loving adults think. According to the available data, the claim that they do not care or will do the opposite is wrong.

We need to understand the message from the incredible $25 million National Longitudinal Study on Adolescent Health (ADD study) reported in the *Journal of the American Medical Association*.[19] This is the largest study ever done on adolescents in the United States. It shows that, above everything else, young people respond to their parents' guidance in avoiding risky behavior. Also highly statistically significant, concerning the issue of sexuality, is the finding that young people who have taken a pledge of abstinence are significantly less likely to engage in risky behaviors than young people who have not. Surprisingly, 15 percent of girls and 10 percent of boys in this national survey had taken such a pledge. We need to encourage our children to take such a pledge and support programs that embrace this approach.

19. M. D. Resnick, P. S. Bearman, R. W. Blum, K. E. Bauman, K. M. Harris, J. Jones, J. Tabor, T. Beuhring, R. E. Sieving, M. Shew, M. Ireland, L. H. Bearinger, and J. R. Udry, "Protecting Adolescents from Harm: Findings from the National Longitudinal Study on Adolescent Health," *Journal of the American Medical Association* 278, no. 10 (1997): 823-32.

Why would not all parents give clear direction to their own children? Apparently many parents have accepted the idea that they cannot significantly influence their teenagers. Having accepted this untruth, many parents then withdraw from providing guidance to their children. We must first educate parents about the powerful evidence from the ADD study that parental influence is the greatest deterrent to risky behavior among adolescents. Medical offices, schools, community meetings, and our churches are all ideal places to educate parents. Op-ed pieces in our newspapers and letters to the editor can also help communicate this message.

Once parents are aware of their potential influence, they must become educated about the nature of the problem. They need to learn about the various statistics already cited. But we also need to teach parents, educators, youth leaders, and others who love children that risky behavior is often a syndrome. If teens take risks in one area of their behavior they are very likely to do the same in other areas (see the chapter by David Hager in this volume).

There are no quick fixes to the problems in view here. In his book *Why Johnny Can't Tell Right From Wrong*, Bill Kilpatrick says that we must take the time to help young people develop strong values and strong character. According to Kilpatrick, we cannot just tell people, "I've worked it out for you to run the Boston marathon," and expect them to run it successfully if they have not developed the physical foundation for running marathons. In the same way, we cannot just say to young people, "don't have sex," unless we have first helped them build the internal values and character for being able to make wise decisions.[20] More specifically, "Just Say No" programs (those that only provide a lecture or two about abstinence from sex) do not work for most young people in helping them avoid risky behavior. A young person's ability to say no to any of the risky behaviors must be based on the foundation of strong character and values. Knowing this, we can use our influence in society, including schools, medical clinics, faith communities, and elsewhere, to institute long-term, character-based sexuality education programs.

Meanwhile, there is a pressing need for people to better understand the condom issue. The fundamental question is, "What will increase the chance that a young person will grow up with the most healthy, happy life possible?" The issue is not condoms — it is not just risk reduction. The issue is elimination of risk. Condoms only reduce the risk for some diseases for a short period of time. Condoms provide almost no protection against the most common STD, which is human papilloma virus; and they provide very unreliable

20. W. Kilpatrick, *Why Johnny Can't Tell Right From Wrong* (New York: Simon and Schuster, 1992).

protection against Chlamydia, which can cause infertility. Still, even if condoms were more effective in preventing pregnancy, and even if more young people used them, too many of our youth would still be getting hurt physically and emotionally by sex.

Another insufficiently understood piece of the sexuality puzzle is secondary virginity. Secondary virginity refers to singles who have had sex in the past and have returned to abstinence. Data clearly show that if young people are sexually active and not married, they will usually experience an increasing number of sexual partners. In fact, *Sex in America* observes that, "The reason that people now have more sexual partners over their lifetime is that they are spending a longer period sexually active, but unmarried."[21] The data also clearly show that the risk for sexually transmitted disease is directly related to the number of lifetime sexual partners. The imperative, therefore, is for medical people and parents especially, but also for entire communities, to help guide young people who have become sexually involved back to remaining abstinent until they are married — if for no other reasons than public and personal health. We need to guide them confidently to secondary virginity as we help them understand that if they remain in the sexually active lifestyle, they are at risk for sexually transmitted infection and non-marital pregnancy.

There are important foundational issues at stake here as well. We can and should teach the truth about the importance of marriage, even in today's multicultural society. Marriage is the environment for sexual activity, for it is the environment in which people are most likely to avoid disease, unsupported pregnancy, and emotional damage. *Sex in America* similarly concludes that marriage "regulates sexual behavior with remarkable precision. No matter what they did before they wed, no matter how many partners they had, the sexual lives of married people are similar. Despite the popular myth that there is a great deal of adultery in marriage, our data and other reliable studies do not find it. Instead, a vast majority are faithful while the marriage is intact."[22] Most sex educators say that they want to give full information about sexuality and sexual issues to young people, but many of these educators do not talk about marriage. Since so many sex acts occur in marriage, and since that is the environment in which people are most able to avoid disease and non-marital pregnancy, the question we must put to many of today's sex educators is: "Why are you afraid to speak of marriage?" Full, honest sex education demands that marriage be discussed, including its benefits. *Sex in America* and

---

21. *Sex in America*, p. 89.
22. *Sex in America*, pp. 88-89.

other reputable studies point out that people who are married not only engage in sex more than sexually active singles, but enjoy it more!

Faith is also important if young people are to avoid risky behavior. The ADD study and a number of other research reports have shown this. The ADD study, for example, states: "Those who ascribed importance to religion and prayer tended to have a later age of sexual debut and were also less likely to use substances. This is consistent with other studies of risk and protective factors that link religiosity, spirituality, and religious identity with conventional behaviors."[23] Many children do take their faith seriously. But whether they have a faith or not, they can learn values.

That sex is a normal human function does not mean it should have no guidelines. Several years ago much press was devoted to the story of a thirty-five-year-old teacher impregnated by her thirteen-year-old student. She was quoted as saying: "When the sexual relationship started, it seemed natural. What didn't seem natural was that there was a law forbidding such a natural thing."[24] Scott Peck, author of *The Road Less Traveled,* speaks very plainly when he says that we are often most human when we do that which is not natural. It is natural for us to defecate in our pants, but we learn not to do so.[25]

It is imperative that we help our young people, and indeed everyone, understand that sex is like fire. It must be handled appropriately. In the right context, sex is a wonderful gift. It allows true intimacy in a context of fidelity. It allows procreation in an environment that enables children to be raised in a healthy fashion, with more hope than when they are born outside of such an environment. Sex handled appropriately provides stability for a culture.

There was an epidemic during the 1800s, the enormous scope of which reminds me of the STD and non-marital pregnancy predicament of today. Medicine failed to protect humanity for many years during that epidemic, although the solution to the problem was available. A key but unheralded figure of the day was an obstetrician named Ignaz Simmelweis.[26] Maternity hospitals in Europe had an ongoing maternal death rate of about 20 percent: 20 women out of 100 who came into hospitals to deliver a baby died. Simmelweis found that if physicians simply washed their hands in a mild antiseptic solution after autopsies and before examining patients, the

23. Resnick et al., "Protecting Adolescents," p. 831.

24. "Boy, 14, Who Fathered Teacher's Child Dons Ring," *Dallas Morning News,* 23 August 1997, p. 6A.

25. M. S. Peck, *The Road Less Traveled* (New York: Simon and Schuster, 1978).

26. K. C. Carter, "Ignaz Simmelweis, Carl Mayrhofer, and the Rise of Germ Theory," *Medical History* 29 (1985): 33-53.

death rate dropped precipitously to approximately 4 percent. Because he was on the outside of the political and medical establishments of his day, his work was rejected throughout his lifetime. It was not until years after his death that society realized he was right and that so many had suffered needlessly.

As in Simmelweis's day, hundreds of thousands are suffering. But their suffering is the disease and death of sexually transmitted disease and the tragedy of non-marital pregnancy. Some steps that could be taken to solve these problems are not currently embraced by many of today's political and medical leaders.

May we have the heart of a Simmelweis to continue to point our culture toward a healthier way forward. If we are faithful and persistent, our voice will be heard and echoed across the land — and we will give birth to the *new* sexual revolution.

CHAPTER 13

Sexually Transmitted Disease, Preventive Public Policy, and Postmoderns: The Spider, the Web, and the Fly

Mary B. Adam, M.D.

Sexually transmitted diseases (STDs) have reached epidemic proportion in the United States. The reported number of cases of genital *Chlamydia trachomatis* infections alone exceeds that of all other noticeable infectious diseases in the United States. Adolescents ages ten to nineteen and young adults ages twenty to twenty-four have higher rates of STDs than adults over twenty-five.[1] Adolescents make up about 10 percent of the population but account for a disproportionate amount of the STDs in the United States Of all new HIV infections, 25 percent are diagnosed in people under age twenty-two.[2] The severe nature of this public health problem has led to the development of strategies designed to stem the tide of STDs. These strategies to date have been ineffective in decreasing the devastating impact of the epidemic.

---

1. The Centers for Disease Control, "Sexually Transmitted Disease Surveillance 1996."
2. S. K. Schonberg and N. D. Hoffman, "Adolescents and HIV: Controversies Old and New," in *Controversies and Conundrums,* ed. S. M. Coupey and R. T. Brown (Philadelphia: Hanley and Belfus, 1997), pp. 463-72.

Inadequate Prevention Strategies

A variety of prevention strategies have been implemented in an attempt to stop the spread of STDs. Case identification and treatment has long been a goal of public health officials. County STD clinics that offer free testing and treatment have been in place for decades. Hopes of STD eradication resulting from these approaches were shattered in the 1980s with the advent of a new and fatal STD that came to be known as AIDS. In the AIDS era, STD prevention policy received special public attention. The Centers for Disease Control and Prevention and then Surgeon General C. Everett Koop aimed a prevention message at high-risk patients, emphasizing condom use for each and every act of sexual intercourse. This message, initially intended for high-risk groups only, was generalized to the whole population. A wide variety of education efforts followed, which were aimed at educating the population as a whole, especially adolescents and young adults.

This condom message has become a mantra for the 1990s and has been universalized to the point that "safe sex" is equated with condoms. The message has infiltrated all segments of our society. When asked about safe sex, adolescents will uniformly regurgitate the "use a condom" message. Actual practice, however, has shown that this prevention strategy is ineffective, as well as poorly and inconsistently used. Adolescents, who are at the greatest risk for acquiring an STD, are far more likely to use a condom for prevention of pregnancy than for protection from an STD.[3] The data on condom effectiveness has shown that while condoms are about 90 percent effective in decreasing HIV transmission,[4] they are completely ineffective in decreasing chlamydia[5] and human papilloma

---

3. D. P. Orr and C. D. Langefeld, "Factors Associated with Condom Use by Sexually Active Male Adolescents at Risk for Sexually Transmitted Disease," *Pediatrics* 91 (1993): 873-79.

4. M. D. Guimaraes, A. Munoz, C. Boschi-Pinto, and E. A. Castilho, "HIV Infection among Female Partners of Seropositive Men in Brazil. Rio de Janeiro Heterosexual Study Group," *American Journal of Epidemiology* 142 (1995): 538-47; I. de Vincenzi, "A Longitudinal Study of Human Immunodeficiency Virus Transmission by Heterosexual Partners. European Study Group on Heterosexual Transmission of HIV," *New England Journal of Medicine* 331 (1994): 341-46; A. Saracco, M. Musicco, A. Nicolosi et al, "Man-to-Woman Sexual Transmission of HIV: Longitudinal Study of 343 Steady Partners of Infected Men," *Journal of Acquired Immune Deficiency Syndrome* 6 (1993): 497-502.

5. M. J. Rosenberg, A. J. Davidson, J. H. Chen, F. N. Judson, and J. M. Douglas, "Barrier Contraceptives and Sexually Transmitted Diseases in Women: A Comparison of Female-Dependent Methods and Condoms," *American Journal of Public Health* 82 (1992): 669-74; J. Pemberton, J. S. McCann, J. D. Mahony, G. MacKenzie, H. Dougan, and I. Hay, "Sociomedical Characteristics of Patients Attending a V.D. Clinic and the Circumstances

virus (HPV) transmission[6] and show only limited effectiveness in decreasing transmission of gonorrhea,[7] herpes simplex virus,[8] and syphilis.[9]

The popularity and persistence of the condom prevention message is disturbing in view of the failure of this message in accomplishing its goals. In fact, this message has remained virtually unchanged since the early 1980s. However, as I will discuss, the broad popularity of this method is consistent with the assumptions about sexuality held by postmodern culture. To understand the profound impact postmodern thought has had on public policy in regard to STD prevention, we will analyze the postmodern view of sexuality and trace its development from the Judeo-Christian and modern views of sexuality. A comparison of recent and previous STD prevention efforts will illuminate the tension between the postmodern and modern views of sexuality. As we dissect the illogical train of thought that has prevailed in recent public policy and reject a "one size fits all" approach, we can offer a multifaceted STD prevention policy that will address the unique needs of specific groups and establish a scientifically sound as well as spiritually redemptive policy.

Views of Sexuality

In order to understand how our culture has been influenced by different views of sexuality, the three most prominent views will be discussed. The Judeo-Christian view of sexuality has had broad influence on our culture. God created man and woman in his own image and stated that his creation was good (Gen. 1:31). Sexual intercourse was ordained by God for both procreation (Gen. 1:26-28) and for the intimacy and closeness of the couple (Gen. 2:18-25). Likewise, the New Testament discusses sexual desire as an ap-

---

of Infection," *British Journal of Venereal Diseases* 48 (1972): 391-96; J. M. Zenilman, C. S. Weisman, A. M. Rompalo et al., "Condom Use to Prevent Incident STDs: The Validity of Self-Reported Condom Use," *Sexually Transmitted Diseases* 22 (1995): 15-21.

6. J. H. Jamison, D. W. Kaplan, R. Hamman, R. Eagar, R. Beach, and J. M. Douglas Jr., "Spectrum of Genital Human Papillomavirus Infection in a Female Adolescent Population," *Sexually Transmitted Diseases* 22 (1995): 236-43.

7. Rosenberg et al., "Barrier Contraceptives"; Pemberton et al., "Sociomedical Characteristics."

8. M. W. Oberle, L. Rosero-Bixby, F. K. Lee, M. Sanchez-Braverman, A. J. Nahmias, and M. E. Guinan, "Herpes Simplex Virus Type 2 Antibodies: High Prevalence in Monogamous Women in Costa Rica," *American Journal of Tropical Medicine and Hygiene* 41 (1989): 224-29.

9. Pemberton et al., "Sociomedical Characteristics."

propriate reason for marriage (1 Cor. 7), and clearly teaches that sexual union is to be confined to husband and wife for life. Adultery and fornication are condemned and chastity is commanded. C. S. Lewis describes chastity in the following manner:

> Chastity is the most unpopular of the Christian virtues. There is no getting away from it: the old Christian rule is, "Either marriage, with complete faithfulness to your partner, or else total abstinence." Now this is so difficult and so contrary to our instincts, that obviously either Christianity is wrong or our sexual instinct, as it now is, has gone wrong. One or the other. Of course being a Christian, I think it is the instinct which has gone wrong.[10]

"Sexual instinct" is viewed differently within a modern framework. The modern era of Western thought has been characterized by trust in human reason and human progress in solving problems. The supernatural has been rejected. Human achievement in science and technology seemed to open up a new age of human possibility. It was hoped that human reason together with objective data from the material world would lead to complete understanding of man and his universe. Systematic thinking and application of the scientific method would ultimately give a complete explanation of the material universe. The material universe was all that existed and God was rejected. The primacy of science and scientific method in the modern view was summarized by Francis Schaeffer: "Modern science was a shift from the concept of the uniformity of natural causes in an open system to the concept of the uniformity of natural causes in a closed system. In the latter view nothing is outside a total cosmic machine; everything which exists is a part of it."[11]

It was based on this closed system, without God, that Darwin developed his theory of the origin of humanity. From his time forward, humanity has been considered a natural phenomenon that has evolved from other earlier life forms. The naturalistic view of humanity led to a naturalistic view of sexuality. Sex has been viewed as a natural human impulse just like hunger. Fulfillment of sexual desire can be accomplished without any feelings of love or specific commitment to one's sexual partner. Sex is required for the propagation of the species, but marriage or lifelong monogamy is not. In modern thought, marriage has some advantage for the raising of children and a loving commitment can increase the pleasure of sex; but neither is required in order to satisfy the natural sexual impulse. For example, many people find them-

10. C. S. Lewis, *Mere Christianity* (New York: Simon & Schuster, 1943), p. 90.
11. F. A. Schaeffer, *How Should We Then Live?* (Westchester, Ill.: Crossway Books, 1976), p. 146.

selves attracted at some time or another to someone for whom they have no feelings other than that of sexual desire. At the same time, this modernist mentality has been cautious about homosexual expression. Homosexuality has been viewed by many as aberrant because it inhibits propagation of the species. Yet, avoidance of STDs acquired through promiscuous sexual activity has generally been considered beneficial for both the individual and the species since STDs can cause harm to offspring and cause possible sterility.

Postmodern thought, on the other hand, can be characterized by a loss of faith in progress and human reason. Systematic thinking and the scientific method are no longer gold standards for understanding the materialistic world. For the postmodern, there are no absolutes and meaning cannot be found in the material world. Individuals must create their own meaning, standards, or rules by their own free choices and deliberate actions. Thus, tolerance and pluralism are highly valued. Conviction is seen as the antithesis of tolerance and pluralism. Cultural mores or judgments regarding sexual expression are viewed as opinions instead of absolute standards. Sex is seen as a private matter that is not open to societal judgment. Repression of sexual freedom for any reason is negative. Tolerance of any individual's idea of sexual expression is seen as a societal good. Such tolerance allows the individual freedom to define his or her own idea of sexuality and the way that it should be expressed. This preeminence of the self means that the independence of the partners is highly valued. Therefore, sex can be enjoyed without any strings attached. Lifelong monogamy is seen as unnatural and multiple sexual partners become the norm. Individual autonomy and tolerance of others' choices are paramount. The prevailing attitude is "Whatever works for you." The availability of contraception divorces the act of sex from reproduction. While there is no harm and likely some benefit in having children, the act of intercourse is no longer about reproduction, and therefore the conception of a child is a separate issue. Consequently, homosexual expression is as valid as any other form of sexual expression and should be tolerated without discrimination.

Postmodern Prevention

The postmodern view of sexuality, with its high view of tolerance, has had a profound impact on current STD prevention policy as seen most clearly in prevention efforts for HIV. Education has been the primary means of HIV prevention. The two elements of this prevention education are (1) the means of viral transmission, and (2) the practice of safe sex (i.e., condom usage).

This "one size fits all" approach, unchanged since the early 1980s, allows for a sufficiently vague strategy so as to prevent criticism of the sexual practices of an individual. To date, HIV educators have preferred a simplified (or simplistic) message which avoids teaching people to make judgments about their sexual partners. This stance led to early support of the policy of not notifying partners of known HIV-infected persons. Thus, protection of individual privacy (i.e., the privacy of the infected HIV individual) has taken priority over protection of the community. This is also illustrated by the emphasis on universal precautions, which include requiring health care workers to use latex gloves whenever handling body fluids. This policy has been proposed in part because disclosure of a patient's HIV status to health care workers would violate the HIV-positive patient's right to privacy. Health care personnel are denied the opportunity to exercise extra caution in certain situations, but instead are expected to always maintain a maximal level of alertness.

Current public policy also protects the autonomy of the individual by not calling for routine HIV testing of pregnant women. Routine testing of pregnant women for syphilis, gonorrhea, and chlamydia are done unless a woman specifically requests that they not be done. To test a woman for HIV requires that she give written consent before any testing — a distinctly different approach.

Another distinctive facet of current prevention policy has been the reluctance to discuss lifelong monogamy, whether homosexual or heterosexual, as an effective strategy. Postmodern sexuality views lifelong monogamy as abnormal. To condone lifelong sexual monogamy as a preferred sexual pattern would require societal judgment. This type of judgment has been avoided despite monogamy's scientifically proven effectiveness in stopping the spread of HIV and all other STDs. The "use a condom" message has been ubiquitous and has not allowed for specific warnings against high-risk sexual encounters like prostitution. This approach is consistent with the nonjudgmental orientation of postmodern thought.

Lydia Temoshok has stated it well:

> HIV educators have continued to insist that people will only understand a simple message — use a condom every time. These educators think that it is better policy to tell people to use a condom with each partner, whether that partner is a spouse or a bar "pick up," than to advise people not to have sex at all with "casual" or anonymous partners. To justify this stance, and to negate the importance of testing and notification as prevention strategies, HIV educators insist that people lie about their past and present sexual lives: a husband or wife will deny an extramarital affair, the person you met

through a shared interest group in your community will say they had an HIV test and that it was negative when in fact they never were tested; an HIV-infected person will not inform you of his/her positive status because of fear of rejection. Because you can't trust anyone, you should just use a condom every time with every partner.[12]

Using a condom can never replace using good judgment in sexual matters. For STDs other than HIV, partner notification and treatment of those individuals has been a mainstay of prevention policy. Routine screening and partner notification of other STDs reflects the fact that prevention policy for those diseases (gonorrhea, chlamydia, syphilis) was developed at a time when rational scientific (modern) thought played a stronger role in public policy. Current policy, which refuses to advocate partner notification for HIV, places excessive value on protection of personal privacy rights at the expense of community health. This protection of personal privacy occurs even though it is well documented that persons with known HIV infection have less-than-stellar honest disclosure rates.[13] The entrenchment of the postmodern view of sexuality has stymied reevaluation of our current public policy in spite of the "evidence." The increasing effectiveness of early multi-drug treatment for those with HIV demands a reconsideration of public policy. So does the effectiveness of single-drug regimens in the prevention both of vertical transmission of HIV from mother to unborn child and of needle-stick injuries to health care personnel. Reevaluation is only now beginning in a few states such as Illinois. However, despite the evidence, the safe-sex message has remained virtually unchanged since the early 1980s. As a result, it has been faulty in several ways.

First, there has been no significant move to educate those with known HIV infection differently from the population at large. Indeed the safe-sex message, "use a condom to prevent getting HIV infection," is no longer a valid or rational concern once one already has the disease. The prevention message to HIV-positive individuals needs to go beyond using condoms and toward emphasizing the responsibility to stop transmission of the virus.

Second, the safe-sex condom message has been taught to young people without a discussion of the facts of condom ineffectiveness in non-HIV STDs. Condoms do not prevent the most common STDs, yet condom effec-

12. "Preventing the Transmission of the Human Immunodeficiency Virus (HIV)," Hearing Before the Subcommittee on Health and Environment of the Committee on Commerce, House of Representatives, 105th Congress, 5 February 1998, serial 105-71, pp. 100-109.

13. M. D. Stein, K. A. Freedberg, L. M. Sullivan et al., "Disclosure of HIV-Positive Status to Partners," *Archives of Internal Medicine* 158 (1998): 253-57.

tiveness in decreasing HIV infection has been generalized inappropriately to other STDs. Adolescents' risk for contracting gonorrhea, herpes, HPV, or chlamydia are all substantially greater than their risk for acquiring HIV. Yet only one prevention strategy is utilized, even when there is inadequate evidence supporting it. This misunderstanding of the facts is common not only among adolescents but also among many physicians (as evidenced frequently in the author's personal and professional experience).

Third, the current prevention strategy provides little to no education on how to minimize one's high-risk exposure. The case of one HIV-positive man allegedly causing an epidemic of fifteen HIV cases (thirteen females, one only thirteen years old — and two of their infants) in rural Chautuaqua, New York, illustrates this problem well. While it is true that had these women chosen to use a condom they might not have become infected, it is equally true that the HIV-prevention message taught in their schools was inappropriate. Women do not use condoms, men do. Common sense requires questioning how many young women are going to be able to persuade a man who doesn't want to use a condom to do so. Will a young woman be sufficiently skilled to put a condom on a man? In light of the evidence that adolescents are poor users of condoms, it seems to make more sense to warn these young women how easily men can hide their sexual history and health status, and to teach them to walk away rather than try to negotiate using a condom.

Fourth, the entrenchment of the postmodern view of sexuality has clouded our ability to duplicate effective and well-publicized prevention efforts utilized in other countries. Thailand developed a nationwide prevention effort that focused the "use a condom" message on one specific high-risk setting: prostitution. This message was publicized extensively and coupled with an intensive effort to screen for and treat STDs in all prostitutes weekly. Brothels that did not adhere to the policy of required condom use were closed. This multimodal and effective strategy violates several basic values of postmodern thought. This policy singled out a certain type of sexual encounter, that of prostitution, and judged it high risk. Thus the policy was judgmental. Individual freedom of choice was limited when brothels were closed for noncompliance. Indeed, coercion was utilized. Scientific evidence (treating non-HIV STDs decreases HIV transmission) was utilized and translated into public policy. Thus, rational scientific methodology was given priority over individual freedom. The successful Thai prevention program does not fit well with the postmodern view, and this type of public policy is not welcome in the United States.[14]

14. W. Rojanapithayakorn and R. Hannenberg, "The 100% Condom Program in Thailand," *AIDS* 10 (1996): 1-7.

Those lured by the deceptive enticement of sexual pleasure without consequences — seemingly possible with condom use — will be entrapped by the dangerous spider of STDs. The spider is aided by the web of public policy. This web, when analyzed, is revealed to be a maze of illogical postmodern thought. If left unchallenged, the web and spider will trap and destroy many innocent individuals.

The reluctance of both scientists and public health officials to reevaluate and aggressively pursue alternative prevention messages is due to the almost universal acceptance of the postmodern view of sexuality. If we are unwilling to reevaluate these basic assumptions, we are unlikely to make additional strides in prevention. However, if one is willing to consider alternate assumptions about human beings and human sexuality, such as those offered by the Judeo-Christian worldview, a measure of hope is rekindled.

Christian Alternatives

The Christian view affirms the dignity of people, who are created in God's image. It views sex as a sacred act given by God for both pleasure and procreation. The clear call to right (lifelong monogamy) is offered not as a method to deny pleasure but as a mechanism to avoid pain. The assumption is that God really does know what is best for his people. Humanity is viewed as capable of making choices for good or evil. Sex outside of marriage is seen as an evil choice — one that produces destructive consequences such as the spread of STDs. The teachings of Jesus are consistent in regard to the handling of those who make bad choices: Love the sinner, hate the sin. Jesus sought to relieve suffering through the many instances in which he healed the sick. He also showed far greater compassion to the woman caught in adultery than the religious leaders and instructed her to go and sin no more.

In following the example of Jesus there is much a biblical Christian can advocate for in the specific area of STD prevention. Five examples here are added to the many other responses advocated in this book.

1. Compassion for the sick can be manifested by strong support for STD screening and treatment. While many STDs are asymptomatic, their effects on the infected individuals and their possible children can be devastating. In a fallen world where sin has a pervasive impact, whatever one can do to eliminate suffering through screening and treatment has some redemptive value.

2. Concern for the innocent can lead to support for efforts at partner notification and treatment, wherever possible, as well as early treatment for HIV infections in spite of the expense.

3. The benefits of lifelong monogamy can be emphasized without shame and with the full support of the medical evidence.

4. The complete truth about condom effectiveness can be presented instead of the current mixed message in which condom effectiveness in HIV prevention is falsely assumed to apply to all other STDs. While condoms have some limited benefit in STD prevention, they offer little protection from a broken heart. Christians can advocate for truth in media presentations and elsewhere in the broader culture.

5. A strategy for influencing public policy on STD prevention can begin with understanding one's own sphere of influence. For instance, a physician can promote a strategy of aggressive screening and treatment for those who are at risk while continuing to teach that there is a way to remove, not just reduce, the risk of STDs. An ethicist can challenge, where appropriate, the arguments of postmodern philosophy with its lack of boundaries and absolutes and the accompanying tendency toward hopelessness and despair. A redemptive approach can always offer hope. A public policymaker can continue to advocate multifaceted prevention messages tailored to specific communities and their unique needs — rejecting the "one size fits all" approach. Parents can teach the importance of God's view of sexuality to their own children.

While these strategies differ, they must have at least one element in common. Everyone must live the chaste lives they call upon others to live.

PART V

OTHER PROACTIVE RESPONSES

CHAPTER 14

Making Laws and Changing Hearts

Gracie Hsu Yu, M.H.S.

The sexual revolution of the 1960s and 1970s in the United States was a revolution of gigantic proportions, for it included attitudinal changes in the scientific, cultural, literary, medical, and political realms. For example, Dr. Alfred Kinsey challenged science to rethink its traditional views about monogamy and children's sexuality when he published his now-discredited tomes on the sexuality of human males and females in the 1950s. Hugh Hefner followed Kinsey's lead in the cultural realm by celebrating promiscuity through his *Playboy* empire. Betty Friedan and Simone de Beauvoir pioneered the idea of female sexual "equality" with their literary works on feminism. Medically, the widespread introduction of "the Pill" into American society made promiscuity more acceptable because it diminished the fears of out-of-wedlock pregnancy. The American Medical Association's sanctioning of birth control and abortion in 1959 and 1967, respectively, also served to give promiscuity the imprimatur of the medical establishment. Finally, the sexual revolution entered the political sphere in 1970 when Congress passed Title X of the Public Health Service Act, which funded contraception for minors and lower-income married and unmarried women. All of these developments gave rise to the sexual revolution, in which people were encouraged to abandon tradition and embrace the making of their own laws with regard to sexual attitudes, relations, and roles.

The ideas and attitudes of the sexual revolution continue to permeate U.S. culture today. Despite the altering of the "free love" mantra to "safe sex," people today still largely abide by the "If it feels good, do it" hedonism that marked the 1960s and 1970s. The disastrous social fallout of that hedonism — illegitimacy, sexually transmitted diseases, abortion, pornography, HIV/

243

AIDS — has, perhaps more than any other reason, caused Christians to recently become active in public policy. Among Christians, Catholics have long distinguished themselves in the fights against abortion and government-funded contraception, even decades before evangelical Christians began to organize politically. But much more needs to be done. How can Christians be more politically effective in combating the destructive trends of the sexual revolution that have become so ingrained in our culture?

Republicans, Democrats, and Christians

The first answer is to recognize that the political realities of Washington, D.C., demand that Christians appeal to both the Democratic and Republican parties, not necessarily in equal measure, but they should at least attempt to understand each party's motivation and find common ground. Recently, Christians have become more of a force among the Republican Party, and their movement has been termed by media pundits as "the religious right." Indeed, exit polls from the 1994 midterm elections that swept Republicans into Congress and statehouses found that religious conservatives played a key role in the GOP victory. For example, a *Washington Post* exit poll found that more than one in four voters identified themselves as "born-again" or "evangelical Christians" and that they voted Republican by large majorities, especially in the South.[1] A Christian Coalition survey of 1,000 voters immediately after the polls closed showed that religious conservatives accounted for 33 percent of all the people who cast a vote, and approximately seven in ten of them voted Republican.[2] And a University of Iowa poll showed that 40 percent of the vote for Republican Terry Branstad in his gubernatorial victory over Democrat Bonnie Campbell came from evangelicals.[3]

While Christians have been active in the Republican Party, the same cannot be said about their involvement with Democrats. Unfortunately, Christian outreach among Democrats, particularly Protestant Christian outreach, has generally been slim to none.

Looking at how Christians might influence both the Democratic and Republican parties illuminates the complex process of good lawmaking, which involves the shaping of public attitudes as well as changing the minds

1. Michael Shanahan, "Religious Right Increases Influence in GOP Gains," *Plain Dealer,* 10 November 1994, p. 19A.

2. Esther Diskin, "Christian Coalition Takes Partial Credit for Voter Turnout, GOP Wins," *Norfolk Virginian-Pilot,* 10 November 1994, p. A14.

3. Shanahan, "Religious Right."

and votes of members of Congress. Indeed, just as the sexual revolution came about through a broader revolution in the cultural, scientific, literary, medical, and political spheres, so should our Christian response be broad in its reach, even venturing outside the sphere of politics.

But let us first examine the varying ways to go about influencing the two political parties. Doing so will help us better understand why concentrating solely on the making of laws falls short.

The political differences between Republicans and Democrats are certainly numerous, but one overarching framework helps explain how each party arrives at the positions they do. In general, Republicans are driven by their heads, and Democrats are led by their hearts. Or, as William R. Mattox Jr. describes it, Republicans are frequently hard-headed and hard-hearted, while Democrats are more often soft-headed and soft-hearted.

Republicans tend to subscribe to a principle they believe to be true and build policy based on that principle. For example, because Republicans believe that human life is sacred, their party platform includes a pro-life plank and their members largely vote pro-life. Because they believe that the U.S. welfare system has fostered illegitimacy, they overhauled the welfare system in order to discourage illegitimacy.

Republicans, however, are also criticized as being hard-hearted regarding the needs of hurting people. For example, when Republicans speak out for the unborn, they often fail to address the needs of the pregnant women who are in crisis. It was this failure that fueled Surgeon General Jocelyn Elders's infamous remarks that pro-lifers should "get over their love affair with the fetus and start supporting the children"[4] and that abortion opponents "love little children as long as they are in someone else's uterus."[5] Similarly, during the debate over welfare reform, Republicans were criticized for leaving the poorest of all Americans to fend for themselves without a government safety net. Though it is admittedly a generalization, Republicans have often been unwilling to compromise on principle, but in the process they have often seemed to lack compassion for those who fall short of their exacting standards.

Democrats, on the other hand, tend to be led first by their hearts. They see the needs of suffering people and build policy to help alleviate that suffering. President Bill Clinton has been a prime example. When Clinton says, "I

---

4. Martin Kasindorf, "Sitting Duck; Elders: An Easy Target for the GOP," *Newsday*, 8 July 1993, p. A34.

5. Christopher Cooper, "Surgeon General Nominee Set to Weather Birth Control Storm," *Times-Picayune*, 24 January 1993, p. B1.

feel your pain," he is appealing to the hearts of the people who are hurting. He demonstrated this soft-heartedness on April 10, 1996, when he vetoed the partial-birth abortion ban bill for the first time. In a White House Rose Garden ceremony, five women told their stories about their agonizing and heart-wrenching decisions to abort their children using the partial-birth abortion technique. Clinton claimed he had to veto the bill so that these five women, and others like them, could continue to make their choice with dignity and respect. Democrats also favor welfare spending because they believe the government must provide a helping hand for the poor and helpless.

At the same time, Democrats are often so driven by their hearts that they fail to acknowledge fundamental principles like the sanctity of all human life. They may also gloss over the real long-term harm that some of their well-meaning programs have caused.

How can understanding this paradigm help Christians reach out to both Democrats and Republicans? Christ's example is instructive at this point: Christians are called to be hard-headed but soft-hearted. John 1:14 says that Jesus came in "grace and truth." Jesus was hard-headed in that he unswervingly abided by God's law and never compromised on the truth. Indeed, he brandished a whip and overthrew the tables of the merchants because they were making the temple a marketplace instead of a house of prayer. But Jesus was also soft-hearted, full of compassion and grace, especially to those who were considered sinners and outcasts in the eyes of society. Luke 4:18-19 records that it is for the poor and oppressed that Jesus came to earth:

> The Spirit of the Lord is on me, because he has anointed me to preach good news to the poor. He has sent me to proclaim freedom for the prisoners and recovery of sight for the blind, to release the oppressed, to proclaim the year of the Lord's favor.

Jesus' characteristic grace was also evident in his interaction with the Samaritan woman at the well in John 4. Although Jesus knew that she was living with a man who was not her husband and that she was drawing water in the heat of the day to avoid the other women's derision, Jesus did not condemn her; but neither did he condone her behavior. He instead spoke to her heart's longing for true fulfillment, telling her that he could give her "living water." She eventually changed her lifestyle because she had come to know that Jesus cared about even her, a Samaritan woman who had sinned against God.

There are many other examples today of how God overwhelms people with his grace in order to bring them to the truth. Norma McCorvey, the

"Jane Roe" of the landmark *Roe v. Wade* Supreme Court decision that legalized abortion, shocked the world when she came forward several years ago to announce that she had become a Christian and had renounced her pro-abortion position. She was convinced to come to the truth through the love of a seven-year-old girl named Emily. The daughter of a woman who regularly protested outside the abortion clinic where Ms. McCorvey worked, Emily told Ms. McCorvey about the horrors of abortion, but she also extended unconditional love and friendship to her.[6]

Ms. McCorvey was softened by the grace of this little girl who also spoke the truth boldly. And when Ms. McCorvey discovered that Emily was almost a victim of abortion herself, she began to realize the truth in a new way. As Ms. McCorvey tells it, it all came to a head one day when she saw a whole row of empty park swings. She looked at these empty park swings and thought to herself, "What happened to all the children?" And then she suddenly thought, "Oh no! They've all been aborted." Somewhere in her subconscious, she knew the hard truth that abortion kills unborn children, but that truth finally came to the surface of her conscious mind at that one moment because she had known the grace of a little girl.

Hard-Headed and Soft-Hearted Public Policy

How can Christians be hard-headed and soft-hearted in the practical world of public policy? How can Christians appeal to both Democrats and Republicans?

To date, Christians have found it easier to be hard-headed in the realm of public policy, advocating laws that are grounded in biblical truth. For example, Christians typically abhor abortion because it is the unjustified taking of an innocent human life. As a result, Christians often find themselves in agreement with Republicans who are generally pro-life for the same reason.

Unfortunately, Christians have also often repeated the same failure as Republicans and proclaimed truth alone without a similar appeal to compassion. Thus, for many years, Christians have lobbied for legislation that would regulate abortion and overturn *Roe v. Wade*, thereby alienating Democrats who are concerned that women will suffer if abortion services are not available.

The Human Life Amendment is a good example of how Christians

6. Norma McCorvey and Gary Thomas, *Won By Love* (Nashville: Thomas Nelson Publishers, 1998).

have often approached pro-life policy. Because abortion is an evil, they reason, the Constitution should be amended to explicitly guarantee the right to life of the unborn. However, the first effort to amend failed in 1984. There is no evidence of increased support for such an amendment in the years since then.

The reason that pro-lifers have been unable to advance their cause through championing a Human Life Amendment is not because such a law is wrong in principle — it is grounded in truth — but because pro-lifers have not persuaded the hearts of people that abortion is not necessary. Indeed, is it likely that people will consider one million abortions a tragedy if they do not consider a single abortion a crime? Besides, if pro-lifers were able to get a Human Life Amendment passed amidst the current tide of public opinion, would people abide by it? Would the police, the courts, and juries evade legal restrictions? Would Congress and state legislatures enforce the amendment's protections? Clearly, pro-lifers' success in passing enforceable laws will ultimately be hindered unless they simultaneously begin to change hearts.

Pro-life Christians can appeal to the hearts of people and attempt to sway public opinion in a variety of ways, including through legislation. One effective piece of legislation that has proclaimed truth and helped influence public opinion is the Partial-Birth Abortion (PBA) Ban Act. Since 1995, Republicans in Congress have sought to outlaw partial-birth abortions, an "abortion in which the person performing the abortion partially vaginally delivers a living fetus before killing the fetus and completing the delivery."[7]

This bill is hard-headed in that it lays bare the truth that abortion kills unborn children and refocuses attention on the plight of the unborn child. But its advantage over the Human Life Amendment is that even the sanitized definition of the partial-birth abortion technique is horrifying. The PBA bill appeals to people's hearts rather than solely to their intellects. Indeed, many people, upon hearing about the PBA procedure, feel as nurse Brenda Pratt Shafer did after she witnessed such an abortion. She wrote, "I think a lot of people think, as I did, of just a blob of cells, or a mass or something. . . . *I don't think about abortion the same way anymore*"[8] (emphasis added).

Because the partial-birth abortion procedure evokes a strong gut-level reaction, recent polls have shown that people are changing their minds about late-term abortions. For example, a January 1998 *New York Times*/CBS poll found that only 15 percent of people in the United States believe that a woman should be permitted to have an abortion during the second three

7. HR 1122, version dated 25 March 1997.
8. Correspondence from Brenda Shafer to Congressman Tony Hall, 9 July 1995.

months of pregnancy, and that only 7 percent believe a woman should be permitted to have an abortion during the last three months of pregnancy.[9] An August 1997 *USA Today*/CNN/Gallup Poll found that 22 percent of Americans said abortion should be legal under any circumstances. This is a marked decrease from 31 percent in September 1995.[10] And "Gallup Polls find 64 percent of adults believe abortion should be legal in the first three months of pregnancy; only 26 percent agree in the second three months. And 71 percent disapprove of the 'partial birth' technique."[11]

The Partial-Birth Abortion Ban Act has also garnered the support of several members of Congress who would normally oppose any type of anti-abortion legislation, including House Minority Leader Richard Gephardt (D-MO), Rep. Patrick Kennedy (D-RI), Rep. James Moran (D-VA), Senate Minority Leader Tom Daschle (D-SD), and Senator Daniel Patrick Moynihan (D-NY), who made the remarkable statement that partial-birth abortions are just "too close to infanticide."[12]

The pro-choice American Medical Association's board of trustees even weighed in, saying in a thirty-five-page report that there are no situations in which "intact dilation and extraction [known as partial-birth abortion] is the only appropriate procedure to induce abortion."[13] The report's conclusion states, "Except in extraordinary circumstances, maternal health factors which demand termination of pregnancy can be accommodated without sacrificing the fetus, and the near certainty of the independent viability of the fetus argues for ending the pregnancy by appropriate delivery."[14] The PBA Ban Act, then, has had a substantial influence on public opinion, congressional members, and the medical establishment, even though President Clinton's veto has twice prevented it from becoming law.

In addition to the Partial-Birth Abortion Ban Act, which appeals to mass public opinion, other pieces of legislation exist that may ultimately help transform the culture by changing the hearts of people on a more individual level. For example, as part of the 1996 welfare reform law, Congress allocated

9. Carey Goldberg and Janet Elder, "Public Backs Abortion, But Wants Limits, Poll Says," *New York Times*, 16 January 1998, p. A1.

10. Richard L. Berke, "G.O.P. Recasts Public Debate over Abortion," *New York Times*, 21 October 1997, p. A1.

11. Kim Painter, "Fueling the Debate: Late Abortions Spark New Controversy," *USA Today*, 11 March 1997, p. 1D.

12. *New York Post*, 3 May 1996.

13. Joyce Price, "No Need Seen For Partial-Birth Abortion," *Washington Times*, 14 May 1997.

14. Ibid.

$50 million for states to fund sexual abstinence education projects. While not every state has used the money according to federal guidelines, several states have put the money to good use, funding programs that comply with the law to promote abstinence until marriage as "the expected standard" of behavior.[15] This money, which will be allocated each year for five years, has heretofore been a missing ingredient in the grassroots efforts to promote abstinence. For years, many solid abstinence programs have struggled because of lack of funding and visibility; this legislation enables many of these programs to finally burgeon.

The impact of the legislation is great. The mere existence of the abstinence legislation has encouraged some states to modify their explicit sex education and to become more open to funding abstinence-centered projects in order to qualify for federal dollars. Most important of all, the abstinence legislation is significantly influencing the lives of young people through the abstinence-centered programs it funds. As these programs successfully help young people resist peer pressure and remain sexually abstinent until marriage, the indulgent culture of the sexual revolution is being transformed to a more chaste counterculture one person at a time.

In a similar vein, Rep. Joseph Pitts (R-PA) is preparing to introduce legislation that would fund pro-life crisis pregnancy centers. Rep. Pitts successfully passed this type of bill in Pennsylvania when he was a state legislator. Like the aforementioned abstinence projects, most crisis pregnancy centers have also operated for years on shoestring budgets and little public recognition. But the work they do is vitally important.

Crisis pregnancy centers (CPCs) help women who feel they have no other choice but to abort their unborn children to choose a better, albeit not easy, alternative. They help a mother choose life for her unborn baby either by aiding her through an adoption process or by providing her with material goods such as maternity clothing and baby necessities. CPCs also provide emotional and spiritual counseling, and some centers provide housing and/or job training for the women.

By helping to fund CPCs, Rep. Pitts's legislation could also transform the culture one person at a time. Not only would such funding extend the compassionate and personalized care offered by CPCs to more women in need, but the visibility brought to the centers by the legislation would also make more women aware that alternatives to abortion do exist. An indirect result of the legislation might also be that the general public develops a more favorable impression of pro-lifers. Emphasizing the work of the centers

15. Public Law 104-193, Sec 912.

would help soften the image of the pro-life movement by demonstrating that pro-lifers care for both the woman and her unborn child.

While these pieces of legislation do not negate federal funding of explicit sex education or pro-abortion related statutes, they do promote solid and compassionate alternatives to explicit sex education and abortion. Some Democrats might be persuaded to support these alternatives now because they provide people with more choices. Other Democrats might eventually support these measures if the programs change public opinion. Finally, the success of these alternative programs brings into question the legitimacy of continued federal funding of family planning clinics, explicit sex education, and abortion-supporting legislation.

More Than Public Policy

Passing laws that uphold Christian values is important, but laws alone are not the answer. While Christians should continue to pursue the passage of good laws, they would be wise not to place too much hope in a political revolution. That is because the very nature of the law is limited. For example, conservative revolutionaries can improve divorce laws to make it harder for people to divorce, but no law can make a man love his wife in the way that he ought. Pro-lifers can pass a law punishing any person other than a parent who transports a minor over state lines in order to obtain an abortion and thereby evade the parental consent laws of the minor's home state. However, there is no guarantee that the child's own parents will not themselves take their daughter to the abortion clinic.

Because of the limited nature of the law, many Christians seek to engage the culture directly, circumventing the lengthy process of promoting and passing legislation. The aforementioned crisis pregnancy centers and sexual abstinence programs are two examples of direct appeals aimed at individuals to transform the culture. These programs have and will continue to operate regardless of federal funding or recognition.

The Southern Baptist Convention's True Love Waits campaign is another effort targeted directly at today's young people to change the culture. Designed to promote sexual purity, the campaign challenges young people to publicly commit themselves to abstinence by signing pledge cards that say, "Believing that true love waits, I make a commitment to God, myself, my family, those I date, my future mate, and my future children to be sexually pure until the day I enter a covenant marriage relationship." In 1994, over 210,000 signed pledges were prominently displayed on the Washington, D.C.,

National Mall — a stunning testimony to the culture that not all young people are "doing it," and a powerful reminder to all abstinent teens that they are not alone in their quest for purity.

In 1995, True Love Waits pledge cards were stacked one on top of the other to so great a height that the pile reached the top of the Atlanta Dome. In more recent years, the campaign has become more decentralized in its operations, preferring to work through local churches and schools, but its impact among young people continues to grow. Even the *Journal of the American Medical Association* has shown that True Love Waits is more than a mere publicity stunt; the campaign has succeeded in changing behavior. The largest national study ever done of adolescents found that these virginity pledges were one of the most consistent predictors of sexual abstinence among young people.[16]

Christians have also used the mass media and direct advertising to appeal to the culture. For example, the Family Research Council launched a campaign in 1994 promoting sexual purity called "Save Sex." One strategy included placing ads in top pop culture magazines such as *Rolling Stone* and *Seventeen*. These full-color ads featured attractive young people speaking a message of sexual purity. For example, the ad copy of one of the five ads entitled "The New Revolution" reads as follows:

> First, they questioned marriage. Said that love should be "free." But "free love" turned out costly. Very costly for some. Now they're pushing condoms. Saying sex should be "safe." But "safe sex" can be risky (to your health and your heart). We think it's time for a new revolution. We think it's time for a love that is real . . . and lasting . . . and pure. A love that sees sex as a celebration of two lives shared together. Forever. That's why we believe in marriage. And why we're saving sex for it.

Two other organizations have used television ads effectively to communicate their message directly to the culture. The DeMoss Foundation sponsored a very popular series a few years ago entitled "Life: What A Beautiful Choice." These ads, filled with pictures of actual babies moving in the womb or children playing on swings, were a soft appeal to pro-life sentiment, intended to remind the viewer of the beauty of life.

The Caring Foundation has taken a different approach, targeting its television commercials not to the general population, but specifically to women of childbearing age, especially to those who consider themselves pro-

16. Michael D. Resnick et al., "Protecting Adolescents From Harm: Findings From the National Longitudinal Study on Adolescent Health," *Journal of the American Medical Association* 278, no. 10 (1997): 823-32.

choice. Through innovative marketing research, the Foundation discovered that the best way to persuade these women is not through argument or debate, but through understanding and addressing their true feelings about abortion. The ads all feature women sharing their own experience, presenting the issue in terms the target audience will understand and find credible.

For example, the research found that women often see abortion as an act of self-preservation, even though they know intuitively that abortion is wrong. The following ad speaks to these fears about self-preservation and depicts carrying the pregnancy to term as an act of strength:

> [A young woman is jogging through city streets. It is raining. As she runs, her inner thoughts are made audible.] "Everyone's telling me how I should feel. . . . It's not like I planned to get pregnant. Not now. [Referring to angry boyfriend, shown in brief flashback:] Telling me how to feel, what to do, then not sticking around when it really counts. So now it's all up to me. But abortion? Not me. I have to live with myself. [Pause. She runs into the distance, skies clearing.] We'll make it. Yeah, we'll make it just fine."[17]

Another ad speaks to the woman's instinctive knowledge that abortion is wrong. Featuring a ballet dancer, the ad talks, in very poetic terms, about "female intuition," the still, small voice inside each one of us that Christians refer to as God-given conscience. The ad gently encourages women to listen to their intuition rather than the voice of convenience by concluding with the following thought: "Your intuition doesn't always tell you what you want to hear, but when you think about it, when was the last time that voice was wrong?"[18]

The Caring Foundation ads are accomplishing exactly what this chapter has been advocating — appealing to the hearts of people in order to change their minds and behavior. And data show that the commercials are indeed changing hearts and actions. In Missouri, the state in which these commercials first aired, the abortion rate dropped 29 percent from 1988 to 1992, nearly six times the national average decline of 5 percent.[19] Michigan and Wisconsin have also been airing the ads for a number of years, and both states have seen a drop in abortions of just under 40 percent.[20] The success of these commercials indicates that Christians can be effective in persuading the culture to abide by moral principles if they are wise enough to appeal to the heart.

17. Paul Swope, "Abortion: A Failure to Communicate," *First Things,* April 1998.
18. Joseph Esposito, "Awakening the Voice of Conscience," *National Catholic Register* 74, no. 27 (1998).
19. Swope, "Abortion: A Failure to Communicate."
20. Ibid.

The Need for Hope

The time is right for Christians to become proactive because there are so many people who are restless for truth and real hope. For example, many young people today have grown disillusioned with the promises of "safe sex" and are instead choosing to champion sexual abstinence until marriage. This countercultural movement is growing to such a degree that even the popular media have taken notice. In the last several years, *Vogue* magazine reported on the new wave of "smart, sassy, and hip" virgins in an article entitled "Like a Virgin, Again";[21] *Newsweek* devoted ample coverage to the abstinence movement with its stories on "Virgin Cool";[22] and the front page of the *New York Times* Style section splashed the headline, "Proud to Be A Virgin: Nowadays, You Can Be Respected Even If You Don't Do It."[23]

Young people are also disillusioned with the culture of divorce. According to a May 1995 CBS News/*New York Times* poll, young people eighteen to twenty-nine years old were more likely than any other age group to oppose divorce as a solution when "a marriage isn't working out."[24] The strong sentiment against divorce possessed by these Generation Xers may be due to the fact that they are also more likely than any other adults to be children of divorce and to have personally experienced divorce's devastating effects.

People are also awakening to the harsh realities of abortion. Rather than experiencing freedom, as abortion advocates had promised, many women who have had abortions suffer from lasting emotional trauma and grief. A Wirthlin Worldwide poll commissioned by the Family Research Council in January 1998 found that most people in the United States (78 percent) strongly agree that "women who have had abortions experience emotional trauma, such as grief and regret."[25] Moreover, post-abortion grief is not limited to women. Phil McCombs, a staff writer for the *Washington Post,* has written an extraordinary first-person article about his abortion experience entitled, "Remembering Thomas." He writes,

> For some instinctual reason, or just imaginatively, I've come to believe that it was a boy, a son whom I wanted killed because, at the time, his existence

---

21. Anne Taylor Fleming, "Like A Virgin, Again," *Vogue,* February 1995, pp. 68-72.
22. Michelle Ingrassia et al., "Virgin Cool," *Newsweek,* 17 October 1994.
23. Judith Newman, "Proud to Be A Virgin," *New York Times,* 19 June 1994.
24. *The American Enterprise,* July/August 1995, p. 104.
25. Commissioned by the Family Research Council, Wirthlin Worldwide conducted a telephone poll of 1,002 American adults 18 years of age or older from 9–11 January 1998 The margin of error is 3.09 percent.

would have inconvenienced me. . . . His name, which is carved on my heart, was Thomas.

I feel like a murderer — which isn't to say that I blame anyone else, or think anyone else is a murderer. It's just the way I feel, and all the rationalizations in the world haven't changed this. I still grieve for little Thomas. It is an ocean of grief.[26]

Like the woman at the well who longed for living water, people today are longing to replace the despair of the Sexual Revelation with true freedom and real hope. If Christians would follow the example of Jesus and appeal to the hearts, not just the minds, of people, the culture may very well respond. Eileen C. Marx, former communications director for Cardinal James A. Hickey of Washington and now a columnist for Catholic publications, helped Phil McCombs find solace from his grief through a healing process of recognition, grieving, and ultimately forgiveness. Other Christians can help transform the culture by using language and imagery that attracts rather than alienates, by offering hope rather than condemnation. By being hard-headed and soft-hearted in all things, whether legislation, television commercials, or crisis pregnancy centers, Christians can begin to successfully reach the very people who most need a message of hope and freedom in the wake of the devastation of the sexual revolution.

26. Phil McCombs, "Remembering Thomas," *Washington Post,* 3 February 1995.

CHAPTER 15

Banning Human Cloning

Charlene Q. Kalebic, J.D.

The crucial cloning breakthrough has occurred. Dolly the sheep has been produced through cloning.[1] Scientists agree that this technology, within the next few years, might be used to clone human beings.[2] The questions that arise from this theological advance are many. Even if we can clone humans, should we? Do people have a legal right to reproduce themselves by cloning? This essay will focus on the United States as one illustrative setting for addressing that question. What legislation surrounding cloning, if any, should states and Congress be able to enact that would withstand a constitutional challenge?

Proponents of cloning argue that it would be beneficial in various ways. For example, they maintain that it would be a boon to homosexual couples and infertile heterosexual couples who want to reproduce.[3] Others claim a presumptive primacy of procreative liberty, advocating a right to procreate by nearly any means desired.[4]

1. Ian Wilmut et al., "Viable Offspring Derived from Fetal and Adult Mammalian Cells," *Nature* 385 (1997): 810. Wilmut, an internationally known embryologist at the Roslin Institute in Edinburgh, Scotland, is backed by PPL Therapeutics P.L.C., a pharmacological company in Edinburgh. See also Sharon Begley, "Little Lamb, Who Made Thee?" *Newsweek,* 10 March 1997, pp. 53, 56.

2. Begley, "Little Lamb," p. 55: "Cloning humans from adults' tissues is likely to be achievable any time from one to ten years from now" (quoting an editor of *Nature*).

3. Anita Manning, "Pressing a 'Right' to Clone Humans," *USA Today,* 6 March 1997, p. 1d. As Randolfe Wicker, the founder of the newly formed Clone Rights United Front, has put it, "[W]e're defending people's reproductive rights. . . . [M]y clone would be my identical twin, and my identical twin has a right to be born." *Id.*

4. John A. Robertson, "Liberalism and the Limits of Procreative Liberty: A Response

256

However, there is significant concern regarding the use of human cloning. The ethical issues are glaring (see a fuller description of cloning and analysis of the ethical issues in John Kilner's chapter earlier in this volume). Among other considerations, there are questions about the health and safety of the cloned child, about individualism and human dignity, about eugenics and mass cloning, about potential abuse including harvesting body parts, and about cloning's effect on the family and social structure. Because of these concerns, several states and Congress are considering legislation that would ban human cloning.[5]

The first issue this chapter discusses is whether the proposed state and federal legislation to regulate and ban human cloning is constitutional. The reasoning the Supreme Court may utilize is examined, as is the most likely outcome should the proposed bans become law and a constitutional challenge arises. This analysis finds that there does not exist a constitutionally protected fundamental right to clone. More specifically, according to the rational relationship test, there are legitimate state concerns regarding the health and safety of clones/children and society. Legislation banning cloning is a rational means to address those concerns and is therefore constitutional. In the unlikely event that the Supreme Court determines cloning is a fundamental right, the compelling interest test suggests that a ban on cloning would be constitutional as long as the ban is narrowly drawn to reflect the government's compelling interests.

The second issue this chapter discusses is the type of regulation lawmakers should enact. Such regulation should include a permanent federal ban on all human cloning or efforts to clone a human, whether or not implantation into a woman's womb is intended. The ban should not include the banning of animal, plant, or tissue cloning or gene research that does not involve the entire human embryo.

---

to My Critics," *Wash. & Lee L. Rev.* 52 (1995): 233; Radhika Rao, "Constitutional Misconceptions," *Mich. L. Rev.* 93 (1995): 1473; Gilbert Meilaender, "Products of the Will: Robertson's Children of Choice," *Wash. & Lee L. Rev.* 52 (1995): 173; Ann MacLean Massie, "Regulating Choice: A Constitutional Law Response to Professor John A. Robertson's Children of Choice," *Wash. & Lee L. Rev.* 52 (1995): 135. According to Duane R. Valz ("Book Review of Children of Choice: Freedom and the New Reproductive Technologies," *High Tech. L. J.* 10 [1995]: 208), John Robertson advocates the primacy of procreative rights based on the notion that they are negative rights against state interference. He claims that these rights extend to genetic selection of offspring characteristics, including the right to abort for gender characteristics.

5. They include Alabama, Arkansas, California, Connecticut, Illinois, Louisiana, Massachusetts, Michigan, New Jersey, New York, North Carolina, Ohio, Oregon, South Carolina, Virginia, and West Virginia.

CHARLENE D. KALEBIC

Constitutional Background

Cloning Is Not a Fundamental Right

The Fourteenth Amendment to the U.S. Constitution prohibits the states from depriving any person of "life, liberty, or property, without due process of law" (the due process clause). The Fifth Amendment has been interpreted to apply the same standard to the federal government. Two tests have been applied to evaluate governmental action: (1) strict scrutiny and (2) rational relationship.

If a law infringes upon a fundamental right the Court will apply strict scrutiny, the highest standard of review. The government must show a compelling interest in order to justify its action. Even if the government can demonstrate that its regulation furthers a compelling interest, it must also prove that the law creates the least possible burden on the fundamental right. If there is a less burdensome alternative, the law will be found unconstitutional. The strict scrutiny standard is exceedingly difficult to meet.

The rational relationship test is much less stringent. If the law does not infringe upon a fundamental right — for example, regulations on business, economic, or welfare matters — the state needs only to show that the regulation bears a "rational relationship" to furthering the government's legitimate interests or concerns.

What is a fundamental right? Those rights listed in the Constitution are protected, such as those found in the Bill of Rights. However, the Supreme Court has also found the highest level of constitutional protection for other rights that are not explicitly affirmed in the Constitution. It has done so through the "due process clause" of the Fifth and Fourteenth Amendments. Using the doctrine called "substantive due process," the Court has found that certain rights are encompassed by the term "liberty" in the due process clause. These rights are "fundamental" if they are so "rooted . . . as to be ranked fundamental." The judges must look to (a) "the traditions and collective conscience of our people," (b) "[t]he entire fabric of the Constitution and the purposes that clearly underlie its specific guarantees," and (c) "those basic values implicit in the concept of ordered liberty."[6] Thus, fundamental but not explicitly affirmed rights are those found in our nation's history or traditions, or are implicit in the concept of "ordered liberty."[7]

---

6. *Roe v. Wade*, 410 U.S. 113, 152 (1973).
7. *Griswold v. Connecticut*, 381 U.S. 479, 493, 495, 500; *Snyder v. Commissioner of Massachusetts*, 291 U.S. 97, 105; *Palko v. Connecticut*, 302 U.S. 319, 325.

The Court has found that the Constitution creates a "zone of privacy" surrounding an individual's private affairs that is free from unwarranted government interference. The Court has interpreted this fundamental right to privacy to include the right of a married couple to use birth control devices,[8] the right of parents to rear and educate their children,[9] a woman's right to terminate her pregnancy by abortion,[10] the right to marry the person of one's choice,[11] and the right not to be sterilized involuntarily.[12] Moreover, the right of privacy thus encompasses important life decisions that are intimate and personal, such as those relating to marriage, procreation, and family relationships.

If the right to clone is found to be within this "zone of privacy," any regulation must meet the highest standard of review. Rarely do these laws pass constitutional muster. One exception is the Hyde Amendment, a federal ban on taxpayer-funded abortions, which was upheld in *Harris v. McRae.*[13] Cloning should be another.

Although activities relating to family and procreation fall within the zone of privacy, critics of cloning argue that not all procreative behavior is subject to the heightened protection of the constitutional right of privacy. They note that one's conduct is subject to regulation for the protection of society, even in cases that are related to the protected activities of marriage, procreation, and raising children. For example, anti-polygamy laws have been upheld based on notions of morality and ordered society, while immunization requirements have been upheld based on concern for public health and the health of the child. Similarly, laws prohibiting sexual activities outside of marriage such as fornication, adultery, and sex with minors are valid. In general, although various activities may touch on fundamental liberties, one's conduct is still subject to appropriate restrictions when the welfare of others might be endangered or undermined by one's unrestrained choice.[14]

Many experts are of the opinion that cloning is not procreation or reproduction, it is duplication or replication; therefore, there exists a difference between cloning and normal reproduction not just in degree but in kind.[15] As

8. *Griswold v. Connecticut.*
9. *Meyer v. Nebraska,* 262 U.S. 390 (1923).
10. *Roe v. Wade.*
11. *Loving v. Virginia,* 388 U.S. 1 (1967).
12. *Skinner v. Oklahoma,* 316 U.S. 535 (1942).
13. *Harris v. McRae,* 448 U.S. 297 (1980).
14. Massie, "Regulating Choice," pp. 162-63.
15. George J. Annas, Congressional testimony before the Subcommittee on Public Health and Safety, 12 March 1997. Annas is a professor and chair of the Health Depart-

such, cloning should not be considered a fundamental right. Although the Constitution protects the fundamental right to bear children, this right should not be expanded so far as to guarantee procreation for everyone at any cost by any method. (The Supreme Court of New Jersey has stated that there is a limit to the right of procreation, which is simply the right to bear natural children through sexual intercourse or artificial insemination and no more.)[16]

In light of the concerns surrounding cloning's destructive effect upon families and the individuals within those families, and the fact that cloning requires no family or marriage (in fact a woman could simply clone herself), the Court should take the same approach that it did in *Bowers v. Hardwick*. There, professing a reluctance to expand the concept of fundamental rights, the Court held that homosexual activity has "no connection" to the constitutionally protected activities, and thus is not a fundamental right.[17] Under this approach, because cloning is a new technology never envisioned or imagined in the country's history or traditions, cloning should not, and hopefully would not, be considered a fundamental right.

The Government Has Compelling Interests

If the right to clone is found to be fundamental, then any governmental regulation must serve a compelling interest and must be written in the least restrictive way possible. There are significant interests served by banning or restricting cloning, all of them compelling. These interests, which can be expressed in non-religious terms, include medical and safety concerns of the cloned child, concerns about individualism and human dignity, concerns regarding potential abuses including harvesting body parts, and concerns about the effects on the family and social structure. These and related concerns are elaborated in John Kilner's ethical analysis earlier in the present volume. Because these concerns are compelling, they override any individual's interest in human cloning. Alternatively, if the right to clone is not held to be a fundamental right, these state interests would nevertheless meet the less restrictive

---

ment at Boston University School of Public Health and founder of the Law, Medicine, and Ethics Program.

16. *In re Baby M*, 109 N.J. 396, 448 (1988).

17. *Bowers*, 478 U.S. at 191. The Court should apply the same language to cloning as in this case, when it held that "to claim that a right to engage in such conduct is 'deeply rooted in this nation's history and tradition' or 'implicit in the concept of ordered liberty' is, at best, facetious."

standard of review (the rational relationship test), because they are rationally related to the advancement of a legitimate governmental purpose.

Medical Concerns

Perhaps the most important universal concern of cloning centers around the potential biological and medical problems of the produced clone/child.[18] In cloning, the DNA from a single cell (in Dolly's case a mammary cell) is used to produce an entire individual. One problem is that the particular cell being cloned may have undergone a mutation.[19] If the gene involved is dormant because the cell has specialized and the gene is now unrelated to the function of the cell being cloned, the mutation may well go undetected. If the lab were unfortunate enough to choose that particular cell to clone, the baby would be born with horrible or fatal defects. Such defects were so rampant among the other sheep clones produced with Dolly that the U.S. National Bioethics Advisory Committee (NBAC) unanimously concluded: "the significant risks to the fetus and physical well being of a child created by . . . cloning outweigh arguable beneficial uses of the technique."[20] Certainly the philosophy to "first do no harm" as stated in the Hippocratic Oath should guide our path.

Experts note that another problem lies in the lack of diversity that would be created in the gene pool by widespread cloning.[21] The clone would inherit all the genetic diseases of the original. Natural reproduction involving individuals from different families does much to ensure that genetic diseases are not passed from generation to generation. This is one reason for the prohibitions of incest and marriage between close family members. Widespread cloning would weaken one of the human race's important genetic protections.

Genetic evolution of the human race also ensures that we win the war against germs.[22] Germs and other microbial enemies continually evolve to find ways to defeat our immune systems and invade our cells. Our counter-

18. National Bioethics Advisory Committee (NBAC), *Cloning Human Beings,* 1997, p. 26. The NBAC was commissioned by President Clinton on 24 February 1997 to investigate the issues surrounding cloning and report back with a recommendation within ninety days. Their report was issued 9 June 1997.

19. Begley, "Little Lamb," p. 59.

20. NBAC, p. 65.

21. Interview with Larry Singer, Professor of Ethics, Loyola University Law School, in Chicago, Ill., 25 March 1997.

22. David Stipp, "The Real Biotechnology Revolution," *Fortune,* 31 March 1997, p. 55.

measure is to mingle male and female genes and create offspring with novel DNA combinations. Microbes equipped to invade one generation's cells find that the cellular locks have changed on the next, thereby defeating the microscopic enemies. Without the genetic evolution produced from both male and female genes, humans and animals would be feeding grounds for germs and epidemics.

Human Dignity

A primary guardian of human dignity in health care is informed consent.[23] Cloning would present ethical problems analogous to those encountered in past experiments on unknowing mentally incompetent people, in that cloning experiments would jeopardize the lives of countless clones without their consent. In addition, ethical concerns would arise similar to those in recent cases involving frozen fertilized eggs where the couple later divorces. In these cases, the courts have held that there is a right not to become a parent against one's will.[24] Furthermore, cloning children who cannot give their consent is ethically problematic.[25] For instance, parents who clone a dying or deceased child would not be reproducing themselves, they would be replicating their child.[26] It is impossible for any child, much less a dead or dying child (the genetic parent in this case), to give informed consent to procreate.[27] In particular, requiring a child to reproduce with his or her own mother (who would most likely provide the host womb) is a bridge not to be crossed.[28]

23. Stephen A. Newman, "Human Cloning and the Family: A Reflection on Cloning Existing Children," *N.Y. L. Sch. J. Hum. Rts.* 13 (1997): 523, 525-30.

24. *Davis v. Davis,* 842 S.W.2d 588 (1992). This case held that awarding custody of the embryos to the former wife for implantation would violate the former husband's constitutionally protected right not to beget a child, and then awarded the former spouses joint custody of the embryos.

25. Newman, "Human Cloning," pp. 528, 529; Annas, Congressional testimony, note 50.

26. Annas, Congressional testimony.

27. Annas, Congressional testimony: "No one should have such dominion over a child (even a dead or dying child) as to be permitted to use its genes to create the child's child. Humans have a basic right not to reproduce. . . . Ethical human reproduction properly requires the voluntary participation of the genetic parents. Such voluntary participation is not possible for a young child."

28. Newman, "Human Cloning," p. 529. "A child cannot take responsibility for the act of reproduction, and should not be burdened with the knowledge that he has been used for such a purpose. Involving the child, however much or little, in a procreative act with its own mother is to cross a barrier between mother and child that must remain absolute and unbreachable."

Human dignity is also at stake in that clones may be viewed as second-class citizens.[29] U.S. society places a great deal of value in the uniqueness of every individual and the sacredness of human life,[30] influenced by Kantian,[31] Christian, and other traditions.[32] There is legitimate concern that the clone/child would be seen as less than a true individual, brought into this world only to serve, or substitute for, the original.[33]

As Nigel Cameron and Gilbert Meilaender explained in their chapters earlier in the present volume and elsewhere,[34] cloning would demean the child by replacing the unique gift of a child with the duplicated production of a clone. In particular, treating the clone as a replacement would have a lifelong detrimental effect on the self-esteem and self-worth of the child, while prejudice from others who may regard the child as abnormal is likely. One can analogize the effect of cloning to the prejudicial effect of segregated schools on African-American children as described in *Brown v. Board of Education*.[35] The cumulative effect of widespread cloning could be the creation of a class of citizens with a sense of inferiority, affecting their mental and emotional development, and producing a detrimental effect on the nation as a whole.[36]

Potential Abuses

A third type of concern is the predictable range of abuses that could easily attend human cloning. For example, cloning raises the specter of a eugenics

---

29. NBAC, p. 66.

30. Jay Katz, quoted in Mona S. Amer, "Breaking the Mold: Human Embryo Cloning and its Implications for a Right to Individuality," *UCLA L. Rev.* 43 (1996): 1659, 1678: "Belief in the idea of individual freedom is a cornerstone of the Western concept of man and society. The common law nurtures and protects individual freedom through the doctrine of self-determination, which confers on each person the right to pursue his own end in his own way."

31. Emmanuel Kant's influence in the present context is discussed in Axel Kahn, "Clone Mammals . . . Clone Man?" *Nature* 386 (1987): 119.

32. Pope John Paul II, "Instruction on Respect for Human Life in Its Origin and on the Dignity of Procreation," *Origins* 16, no. 40 (1987): 697; see also Cardinal Joseph Bernadin, "Science and the Creation of Life," *Origins* 17, no. 2 (1987): 21.

33. Annas, Congressional testimony.

34. Meilaender, Testimony before the NBAC (see NBAC).

35. *Brown v. Board of Education,* 347 U.S. 483 (1954).

36. Newman, "Human Cloning," p. 528: "[W]e cannot help but see a threat to individuality, to a sense of one's own uniqueness, to a sense of self-worth, to an independent, well-integrated personality. We would be tampering with the basic building blocks of both physical and mental life, a social experiment that burdens and puts at risk the existing children and all of their replicas."

ment made possible by the ability to produce children with predictable
ic codes.[37] The atrocities brought about by Nazis in World War II in the
name of eugenics have given us good reason to be worried about this abuse. A
similar social nightmare would be the use of cloning to create work or military forces to achieve certain goals of a society or its leaders.[38] This scenario
appears most plausible in countries where individual rights are not well protected or where dictators rule.

Other concerns of abuse center around the use of cloning simply to
harvest body parts for the original.[39] This kind of abuse is particularly dis-
concerting when lawyers and bioethicists argue that this is one of the "merits"
of cloning.[40] Harvesting one-of-a-kind organs from the clone, such as a heart,
would amount to murder, and there are certainly laws to prevent murder.
However, is the making of a clone simply to donate a kidney or an eye or bone
marrow for the original an adequate reason to clone? At the present time,
there are no laws that limit childbearing based on the reason a couple decides
to procreate.[41] There have been instances, such as the famous Ayala sisters,
where couples have had a second child to donate bone marrow for the first
child.[42] The second child has been allowed to donate, even though a minor
and supposedly unable to give informed consent, on the theory that it would
be in the best interest of the second child to have an older sibling.[43] Many do
not agree with this practice because it treats the second child as a "product."[44]
However, others point out that there are certainly worse reasons for having
children than to save another life. At best, this point is controversial.[45] In any
case, there are better experimental avenues for developing genetically suitable
body materials for transplant. For example, Organogenesis, a Massachusetts

37. NBAC, p. 74.

38. NBAC, p. 69.

39. NBAC, p. 56.

40. John A. Robertson, "Human Cloning," *A.B.A.J.,* May 1997, pp. 80, 81.

41. *Carey v. Population Serv. Intl.,* 431 U.S. 678, 685 (1977). ("The decision whether
or not to beget or bear a child is at the very heart of this cluster of constitutionally protected choices.")

42. Anastasia Toufexis, "Creating a Child to Save Another," *Time,* 5 March 1990,
p. 56; Lance Morrow, "When One Body Can Save Another," *Time,* 17 June 1991, p. 54. (The
Ayala family and the Curry family both conceived children to be a bone marrow transplant
for an older sibling.)

43. Interview with Professor Shewchuck, Professor of Bioethics, Loyola University
Law School, Chicago, Ill., 8 April 1997.

44. Diane M. Gianelli, "Bearing a Donor? Ethical Concerns Raised over Having a
Baby for Marrow Match," *American Medical News,* 2 March 1990, p. 3.

45. NBAC, p. 55.

biotechnology company, has already cloned skin, and is working on cloning other organs, including the ear, liver, and pancreas.[46] Stem cell research is also particularly promising in this regard.

Abuses of cloning are all the more probable because the costs of cloning are relatively low, and the equipment is easily available. Although federal funds in the United States may not be used for research on human fetuses, fetal tissue, and embryos, there are no restrictions on privately funded research.[47] The research costs to produce Dolly were a mere $750,000.[48] The fear is that this type of research or experimentation on human embryos could literally be done in someone's basement, completely undetected, for about $100,000 in equipment with the services of a biology graduate student.[49]

Family and Society

The last major type of concern involves the effect that cloning would have on the social structure, particularly the family.[50] First, cloning would allow for and foster families without men, because cloning removes the need for any male involvement in the reproductive process.[51]

As Charles Sell discusses in the next chapter, fathers are an important part of families.[52] In the inner cities where fathers are often absent, family ties are weak, poverty and crime is overwhelming, and the children have become super-predators at younger and younger ages with no sense of right or wrong or community.[53] Many experts agree that the huge increase in teenage crime

46. *Good Morning America*, NBC television broadcast, 22 April 1997.

47. 42 U.S.C.A. sec. 289g. *et seq.;* see also "Statement of the President on NIH Recommendations Regarding Human Embryo Research," U.S. Newswire, 2 December 1994; P.L. 104-91; P.L.104-208.

48. Larry Reibstein and Gregory Beals, "A Cloned Chop, Anyone?" *Newsweek,* 10 March 1997, p. 58.

49. Wray Herbert et al., "The World After Cloning," *U.S. News and World Report,* 10 March 1997, p. 60.

50. NBAC, p. 70.

51. *Newsweek,* 10 March 1997, p. 21.

52. Steve Berg, "Begotten, Not Made?" *Star-Tribune,* 26 April 1997, p. 7B, citing Lisa Cahill, Professor of Theology at Boston College, testimony before the NBAC; Wade D. Horn, "Why There Is No Substitute for Parents," *Imprimis* 6 June 1997, pp. 1, 3.; John Witte Jr., "Consulting a Living Tradition: Christian Heritage of Marriage and Family," *Christian Century,* 13 November 1996, pp. 1108, 1111.

53. Adam Walinsky, "The Crisis of Public Order," *The Atlantic Monthly,* July 1995. One of the more tragic incidents occurred in Chicago in 1994, when two ten-year-old boys killed a five-year-old boy by throwing him out the fourteenth floor window because he wouldn't steal candy from the drugstore for them. They became America's youngest in-

is largely attributable to the absence of fathers. It is estimated that 75 percent of children without fathers will experience poverty before the age of eleven, compared with only 20 percent of those raised by two parents.[54] Fatherless children are far more likely to be expelled or drop out of school, develop emotional or behavioral problems, commit suicide, and become victims of abuse or neglect. Male children without fathers are far more likely to become violent criminals, evidenced by the fact that a huge 70 percent of current male prisoners serving long-term sentences came from fatherless households. Female children without fathers are more likely to become rebellious and promiscuous.

Cloning weakens the important institution of the family in other ways as well.[55] Legally sanctioned cloning undermines the traditional family by constructing a "family" of individuals connected primarily by contract.[56] Currently, the legal obligation of parents to support and care for their children is based on their biological tie to their children. While biological bonds are irrevocable, contracts can be breached. Therefore, when families come together based on voluntary contracts, there is a greater danger that the parents' sense of duty will be limited to and contingent upon the performance of the contract. Such an arrangement encourages the attitude that family relationships can be easily entered and exited, accepted or rejected.

Thus, with cloning will come the same problems regarding custody and parental rights and responsibilities that are now associated with surrogacy contracts.[57] The Baby M suit is a case in point, in which the surrogate mother

---

mates when they were sentenced to prison, where one was later involved in a gang rape. See Gary Marx, "Boy in CHA Killing Charged in Prison Assault," *Chicago Tribune*, 13 May 1997, p. A1.

54. Horn, "No Substitute," p. 2.

55. Karen Rothenberg, Congressional testimony before the Subcommittee on Public Health and Safety, 12 March 1997. See also generally Radhika Rao, "Assisted Reproductive Technologies and the Threat to the Traditional Family," 47 *Hastings L. J.* 47 (1996): 951, 962-963; Horn, "No Substitute," pp. 1-4; Sara McLanahan and Gary Sandefur, *Growing up with a Single Parent* (Cambridge, Mass.: Harvard University Press, 1994).

56. Rao, "Assisted Reproductive Technologies," pp. 959-65 and n. 45 (noting that assisted reproductive technologies expose the fact that families made with this technology are not biologically determined, but rather are socially constructed by the state, and that biological fathers are legally obligated to support their offspring even when conception occurred without their knowledge or consent).

57. For example, if a male individual wanted to clone himself, he would need a woman to bear his child. This would involve some sort of surrogate relationship between the individual and the birth mother. A surrogate would also be needed for a couple in which the woman was unable to conceive and carry a child, such as the Baby M case. *In re Baby M*, 109 N.J. 396 (1988).

breached her contracts with the intended parents and filed suit to become legal mother of the baby. Although having too many parents, as did Baby M, is a serious problem, an even worse problem arises when the intended parents are the ones breaching the surrogate contract, leaving an innocent child parentless. In at least two cases, a child has been rejected because of medical problems.[58] In a recent California case, the intended father of a child created with donated eggs and sperm and carried to term in a surrogate mother divorced his wife, the intended mother, and repudiated any obligations to the child.[59] In an unprecedented lack of judgment, the trial court declared that the child had no legal parents. (This declaration was later overruled by the appellate court.)

It is easy to imagine more of these types of cases arising if cloning becomes widespread, leaving children with either too many or too few parents. A clone/child could have as many as six parents: the cell donor, the nucleus donor, the gestational mother and father, and the intended mother and father with whom the child would live. A "defective" clone/child could quickly end up with only one parent — or none. The laws have not been able to keep pace with the new kinds of parent/child relationships introduced by reproductive technologies,[60] causing courts to call for some legislative guidance.[61] Cloning would present even more legal difficulties.[62]

Overall, based on the examples set forth by the Court, it can certainly be concluded that there are legitimate state concerns regarding cloning. At the least, these concerns meet the rational relationship test applicable if cloning is not found to be a fundamental human right. Even if it is found to be a fundamental

58. Rao, "Assisted Reproductive Technologies," pp. 965-66 and n. 49. In one case, a child born with a strep infection and microcephaly was rejected by the intended father. Only when a paternity test showed that the surrogate mother's husband was the genetic father did the biological parents agree to accept the child. In another case, when a surrogate mother with a history of drug abuse gave birth to an HIV-positive child, all parties refused custody of the child.

59. Ibid., p. 966 and n. 51. In what experts call the "nightmare case everyone was dreading," John Buzzanca filed for divorce from his wife, Luanne, shortly before his daughter Jaycee was born, and subsequently refused to pay support on the grounds that he was not the girl's legal father. Judy Peres, "Surrogacy Case Breeds New Legal Dilemma," *Chicago Tribune,* 11 September 1997, sec. 1, pp. 1, 22.

60. Peres, "Surrogacy Case."

61. *In re Baby M,* at 469; *Jaycee B. v. Superior Ct.,* 42 Cal. App. 4th 718, 732 (1996). ("Once again, the need for legislation in the surrogacy area is apparent. . . . We reiterate our previous call for legislative action.")

62. George J. Annas, "Human Cloning: Should the United States Legislate Against It?" *ABA Journal,* May 1997, p. 80.

right, under the strict scrutiny test at least some of the above-mentioned concerns — particularly the health and safety of the clone/child, and the right not to procreate in the case of the dead or dying child whom parents wish to clone — fall in the more demanding category of compelling interests. Moreover, a less-restrictive alternative — such as banning cloning to some while allowing it to others — could not survive an equal protection challenge.[63]

Drafting Legislation

If some sort of ban on human cloning is constitutional, then lawmakers must enact appropriate legislation. There are several factors to consider when drafting legislation: whether there should be overriding federal legislation or state-by-state legislation; whether there should be a partial ban or total ban on cloning and cloning research; whether to ban the use of federal funding in cloning, with an appeal to the private sector to follow or to ban all cloning; whether the ban should be temporary or permanent; and whether other less problematic options for childless couples should be encouraged.

Federal Legislation Is Needed

The advantages to federal legislation, as opposed to ad hoc state legislation, are many. Federal laws would ensure national uniformity on an issue that is

63. For example, prohibiting cloning to some, such as non-married egomaniacs whom some may deem unworthy of the technique, while allowing cloning to others, such as infertile married couples in loving long-term relationships who have tried other alternatives, may pose an equal protection clause challenge. Although, the Court has routinely upheld laws protecting and fostering the special treatment of married individuals and the institution of marriage, none of the benefits allowed to married people but denied to non-married people have been found to be a fundamental right (*Bowers v. Hardwick*, 478 U.S. 186 [1986]). For example, tax laws allow for special treatment of married individuals because economic rights are not fundamental (see *Nebbia v. New York*, 291 U.S. 502 [1934]), and therefore economic regulations are held only to a rational relationship standard under an equal protection challenge. See *U.S. Railroad Retirement Bd. v. Fritz*, 449 U.S. 166 (1980). However, the Court has found that the right not to procreate (i.e., the use of contraceptives, abortion) is a fundamental right, and thus has allowed the use of contraceptives and abortion by both married and non-married individuals. See *Carey v. Population Serv. Int'l*, 431 U.S. 678 (1977); *Roe v. Wade*. Therefore, allowing cloning for only married couples as a last alternative only after natural methods, artificial insemination, fertility drugs, and in vitro methods have been attempted would probably not withstand an equal protection clause challenge.

of such magnitude that it literally affects the way in which humanity will survive. Moreover, a federal initiative would override various state efforts to legislate, which are more likely to suffer from ambiguous drafting, penalties so low that they would provide neither a deterrent nor a punishment, and a lack of knowledge of the issues. In addition, federal legislation would prevent "forum-shopping" in which researchers and potential users of clone technology would relocate to states where the protections against dangerous uses are fewer or not as well drafted.[64]

A Complete Ban on Cloning an Entire Human Is Needed

The only acceptable way to handle this incredible but potentially destructive technique is to legislate a complete ban over the use of cloning to form an entire human being. We must ban not only the implantation of a cloned embryo into a woman's womb, but also any technique that would clone an entire human embryo outside the womb. This comprehensive ban is needed to prevent abuses. However, the statute must be carefully worded to allow cloning of DNA sequences, organs, and tissues — which has already resulted in important scientific and medical advances. In addition, cloning research in animals must be allowed where there is potential for advances in animal husbandry and pharmaceuticals to benefit people.

The National Bioethics Advisory Committee (NBAC), at least in part, agrees with this view in their June 1997 Report and Recommendation on cloning. The NBAC does not, however, go far enough. They simply conclude that "somatic cell nuclear transfer and implantation into a woman's body would at this time be an irresponsible, unethical, and unprofessional act" and recommend that federal legislation be enacted to prohibit the "creat[ing] of a child."[65] Theoretically, this would allow researchers to clone an embryo and support the embryo's growth into a fetus in artificial amniotic fluid as long as it is not implanted into a woman's womb. The embryo could then be manipulated, experimented on, and destroyed at will, because the government does not recognize an unborn human as a person protected by the Constitution until the child is born, based on the Supreme Court's decision in *Roe v. Wade*.[66] For reasons ex-

64. NBAC, p. 100.
65. NBAC, pp. iii, iv.
66. *Roe v. Wade*, pp. 133, 158. The Court concluded the word "person" as used in the Fourteenth Amendment does not include the unborn. Therefore, the Court does not recognize the unborn as a person with any of the rights that are accorded to humans under the Constitution.

plained in Robert Evans's chapter earlier in the present volume, the view taken by the Court that unborn children are not persons is vehemently and rightly opposed by many.[67] Legislation must be written carefully to prohibit all cloning techniques that would clone, or lead to the cloning of, an entire human from the fertilization of the human egg to the birth of a cloned human.

The Cloning Ban Must Cover Public and Private Funding

Implicit in the above analysis are the further conclusions that a ban must cover both the private and public sectors and must be permanent. The initial approach to human cloning in the United States has been a moratorium on the use of federal funding in human cloning research and a request of the private sector to voluntarily refrain from human cloning research.[68] However, in light of the enormous financial opportunities for those who would participate in cloning humans, federal legislation is needed to prohibit the use of cloning techniques on humans in all sectors. There is no other way to effectively deter what is sure to be a very profitable practice.

The Ban Should Be Permanent

The issue of the permanence of a ban arises because the technology could exist at some point to clone individuals safely and without the fear of "mistakes" or abnormalities. Should we then allow cloning humans — or at least keep the matter open for discussion, as the NBAC has proposed in its recommended temporary ban? Dr. Edmund Pellegrino explains why such an approach is unacceptable: "First, it suggests that something inherently wrong can be made right, and second, it begs the question of the moral wrong of human embryo experimentation which is the first and essential step in any cloning of human beings. . . . [A] permanent ban should be placed on human cloning because it depends on a first step which is itself morally indefensible."[69] This view is

---

67. Ibid., p. 161; Pope John Paul II, "Instruction on Respect," pp. 700, 701; Bernadin, "Science," p. 24; Edmund D. Pellegrino, testimony before the Congressional Subcommittee on Public Health and Safety of the U.S. Committee on Labor and Human Resources, available in 1997 WESTLAW 332084 (June 17, 1997). Pellegrino is the John Carrol Professor of Medicine and Medical Ethics at Georgetown University.

68. President William J. Clinton, "Remarks by the President on Cloning to the Press," 4 March 1997.

69. Pellegrino, testimony before the Congressional Subcommittee.

shared by an overwhelming majority of the American public. According to a CNN/*Time* poll, 93 percent of the respondents thought cloning was a "bad idea," and 89 percent of the respondents thought that cloning was "not morally acceptable."[70]

*   *   *

This chapter proposes that, in light of all that is at stake in human cloning, there needs to be federal legislation encompassing a complete and permanent ban on human cloning of the entire individual, from fertilization to birth. Such legislation can indeed withstand a constitutional challenge. We must do all that we can to protect the youngest and most innocent of people — our children and our children yet to be born. We must protect them because they cannot protect themselves. We must protect them from a technology that can have devastating effects, even when used with the best of intentions. Children need to be loved for who they are, not ordered from a menu. Having children is not a right to be demanded at all costs, it is a privilege to be cherished. Despite the modern legal philosophy, children truly are a gift from above.

70. Senator Christopher S. Bond, Congressional testimony before the Subcommittee on Public Health and Safety, 12 March 1997 (available on WestLaw in USTESTIMONY file).

CHAPTER 16

Affirming the Family

Charles M. Sell, Th.D.

In a book published in 1984, a group of futurists attempted to predict what marriage and family will be like in the year 2020.[1] By then, they claim, all sexually transmitted diseases including AIDS will be eradicated, thereby removing one of the major objections to free sex. The other obstacle, pregnancies to unwed mothers, will be dealt with by requiring all people to be sterilized during their teenage years, and their sperm and ova placed in banks for use when needed. The governments by that time will realize that their attempts to manage sexual relationships between people have been in vain and are no longer needed. Thus, people will have the freedom to have sex whenever they choose, without any constraints. To protect children, however, the various governments will require a couple to enter into an agreement to remain together for as long as it takes to care for any children they decide to bring into the world. When they so agree, their sperm and eggs will be fertilized and the fetus brought to term in an artificial womb.

One of the authors proudly applauds these developments:

In separating sexual/sensual/erotic intimacy from human reproduction and marriage, we have finally allowed both experiences to become more fully human and personal. We have received that primal "polymorphic perversity" or "panerotic potential" Sigmund Freud and Norman A. Brown wrote of a century ago in such visionary terms.[2]

1. Lester A. Kirkendall and Arthur E. Gravatt, eds., *Marriage and the Family in the Year 2020* (New York: Prometheus Books, 1984).
2. Robert Francoeur, "Transformations in Human Reproduction," in *Marriage and the Family,* pp. 80-105.

Can these predictions possibly be accurate? Actually, some of what these writers predicted in 1984 is already becoming a reality. For example, as if he were already living in the year 2020, Robert Francoeur writes:

> Lesbian couples purchased frozen semen from sperm banks and had their families by artificial insemination. Gay couples hired surrogate mothers who were then inseminated with their mixed semen. Single heterosexual males hired surrogate mothers; single women called on local sperm banks.[3]

In the light of such revolutionary developments in reproductive technology and morality, how should we respond?

Affirming the Importance of Genetic Family Ties

First, we can affirm the importance of genetic family ties. This is a lesson I have learned personally. When my sons were teenagers, I sensed a problem in communicating with them. As an educator, I knew that communication gaps between teens and parents are common. But mine seemed worse. I had a hard time sharing what I was feeling. I felt like I had a "frog in my heart."

I judged that my intimacy problem with my sons was somehow related to my lack of communication with my eighty-year-old father. I had never felt close to my father, a confessed alcoholic who at the time was in recovery. Part of the problem was mine, since I had moved out of town and my visits home were infrequent. In order to close the gap between my sons and me, I tried to cultivate my relationship with my father. When home, I asked him questions about his past in an attempt to make our conversations more personal. Yet, I could not seem to lessen the distance I felt between us.

A breakthrough happened after my sister asked me to see my father when he had been hospitalized and was possibly near death. Traveling to see him, I resolved somehow to get close to him. My brother and I entered the hospital room together. After our initial greetings and discussion of his situation, I summoned the courage to reminisce about our good times as a family — the trips to the circus and amusement parks. In the midst of this, my father said: "I made a fool of myself with that drinking, didn't I?" Shocked to hear him mention something we had never discussed, I briefly answered that it was okay with me. He had never been abusive to me, only kind, and I had harbored no ill feelings against him. Then I asked him — I am not sure why even now — if he ever got depressed. He had always appeared to be such a light-

3. Ibid.

hearted man. "Yes," he replied, "I do." I followed that with "What do you do then?" "I just lie here and talk to Jesus," he answered — for the first time declaring to me that he had a relationship with God. Though conversation that afternoon was brief, it was deep. I floated out of that room, exalting in the closeness that I felt with him. Because my dad did not die that year, but lived several more, we had other moments of intimacy. Once, when I prayed with him, I had the opportunity for the first time to hear him pray.

Often I have asked myself, "Why did those moments with my father mean so much to me?" Perhaps the answer is that the Creator has endowed us with inevitable biological ties. When these relationships are good, we profit from them; when faulty, we may suffer. In a recent study, medical students were asked, "Are you close to your parents?" Years later, among those who had answered "no" there was more mental illness and cancer.[4]

A strong psychological connection between family members may, of course, merely be a cultural phenomenon. Any emotional suffering may be due to failed expectations bred in a social context that values family ties. Yet, many sociologists maintain that biological relationships are special. Sara McLanahan and Gary Sandefur argue that the cause of possible harm to a child growing up without a father is due to the child's biological bond to the father. Biological connection "adds intensity of affection, identification, and investment and, when it is missing something important is lost."[5] Many evolutionary psychologists also identify something distinctive about genetic ties. They have long suggested that parents, grandparents, siblings, aunts, and uncles tend to be more concerned than are other people about the children to whom they are related.[6] This "kin altruism," according to these theorists, is due to the idea that people fight, not merely for their own survival, but also for the survival of those who carry their genes. Whether or not we agree with this explanation, they join the substantial ranks of those who recognize that close genetic relationships are distinct from those more distant.

Aristotle also recognized this when he disagreed with Plato's attempt to wipe out family ties. Plato proposed that civil harmony and true equality would be enhanced if a child were taken from its parents after birth and raised by the state. Parents were to be kept ignorant about which child was ac-

4. Geoffrey Cowley, "Healer of Hearts," *Newsweek*, 16 March 1998, p. 55.

5. Don S. Browning et al., *From Culture Wars to Common Ground: Religion and the American Family Debate* (Louisville: Westminster John Knox Press, 1997), p. 107. Sara McLanahan and Gary Sandefur, *Growing up with a Single Parent* (Cambridge, Mass.: Harvard University Press, 1994), p. 38.

6. William D. Hamilton, "The Genetic Evolution of Social Behavior, II," *Journal of Theoretical Biology* 7 (1964): 17-52.

tually theirs. Thus, people would treat all children as their own, resulting in better care for all of them. Not so, Aristotle argued; rather, this policy would lead to the neglect of all children because that which is common to the greatest number receives the least care. For Aristotle, parental care and concern is due to the recognition that the child is their own.[7]

Throughout the history of the church, arguments from philosophy, natural theology, ethics, and Scripture have been advanced to affirm the importance of our genetic ties.[8] The biblical evidence is particularly convincing. Though not specifically mentioning that emotional harm springs from the disruption of these relationships, Scripture does make it clear that they are both special and significant. Children are to be conceived and raised by two parents who are committed to one another, according to Genesis 1:28 and 2:24. The latter of these verses — "For this reason a man will leave his father and mother and be united to his wife, and they will become one flesh" — describes both the basis of marriage (a commitment implied in the phrase "be united") and its nature ("one flesh"). That the two become "one flesh" indicates that their conjugal relationship is to be like a blood relationship (consanguineous). The flesh and blood relationship becomes the model for the marriage relationship. Accordingly, this verse implies the distinctiveness of the consanguinal family.

Subsequent Old Testament Scripture shows an ongoing emphasis on these ties, described in phrases such as "flesh and blood," and "bone and flesh" (Gen. 29:14; 37:27; Judg. 9:2). In fact, both marital and genetic relationships are morally reinforced in several of the Ten Commandments. For example, husbands and wives are not to commit adultery and children are to honor parents. Moreover, obligations and privileges of these relationships are addressed in Old Testament laws, leaving no doubt that family ties are to be considered distinct.

The New Testament continues that affirmation. Though Jesus strictly demands primary loyalty to himself/God (Matt. 10:37-38), he uncompromisingly stresses the obligation of children to their parents (Mark 7:9-13) as well as the marriage commitment (Matt. 19:1-12). The Epistles reinforce parent-child obligations (Eph. 6:1-4) as well as those of husbands and wives (Eph. 5:22-33; 1 Cor. 7:10-16).

Affirming the distinctiveness of these biological ties is not intended to denigrate adoptive and blended family relationships or suggest they cannot have the quality of biological ones. Nor does this affirmation imply we cannot

7. Browning et al., *Culture Wars*, pp. 116-17.
8. See Browning for a summary of these.

be psychologically whole or healthy if we are alienated from our biological family. Jesus suggests that he himself might even cause a disruption of these ties — something his followers can obviously accept (Matt. 10:34-36). Yet, in general there is usually an emotional and social cost to simply ignoring or rejecting these ties which God has built into the created order of things.

Sadly, genetic family ties are being threatened by a number of factors. One threat arises from particular forms of reproductive technology. Donor insemination and donor-egg in vitro fertilization are, ironically, both affirmations and renunciations of the importance of these ties. The desire of a couple to have a child that is genetically tied to one of them is a recognition of the value of our biological connections. Yet, anonymous donors of sperm or ova typically exercise little concern for the welfare of their genetic material. Is such a "donation" simply an altruistic gesture to make some infertile couples happy, or is it also a gesture of abandonment, relinquishing control and responsibility for one's genetic offspring? "What if," I asked myself, "my sperm were used to produce a son of a Chicago drug lord who was determined to train him to follow his steps?" What responsibility before God do I have for my reproductive materials?

A second threat to genetic family ties is the way we are diminishing the importance of the biological family in our discussions of the definition of family. It seems that the word "family" is losing its connection to the nuclear and extended genetic family.

An extensive national survey of families conducted in 1989 showed this to be the case. Many Americans do not think of the family primarily in structural terms, as a group of people related by blood, marriage or adoption, but in emotional terms, as a group of people who love and care for each other.[9] Most teenagers, according to pollster George Barna, say the word "family" refers to their close friends.[10] It is ironic that the term "family" conjures up thoughts of warmth, intimacy, and caring but is not reserved primarily for the biological relationships which the word originally described.

Even family ministers find it difficult to agree on a definition of the family. Much of the confusion among them, as among others, is due to the presence of so many forms of today's families: single parents, blended family, stepfamilies, couples without children, and grandparents raising grandchildren. Certainly, we should recognize these as genuine families. Though increasingly prevalent today, they have always existed. Moreover, God has dignified adoption by using it as a metaphor for our relationship to him.

9. *The Future of the American Family* (Chicago: Moody Press, 1993), pp. 27-30.
10. Ibid.

However, current variations of family should not distract us from a strong affirmation of the biological family. It is this family that is a reference point for all the others. It is important to recognize that the others are different from the biological family and, in some cases, specially challenged. The Old Testament emphasizes that the fatherless and widows are often in need of community support. The prophets frequently castigate the nation of Israel for neglecting them. The apostle James claims that true piety involves caring for the fatherless and widows *in their distress* (James 1:27, emphasis added).

David Blankenhorn has reviewed the research which shows how single-parent families are challenged and will fail to socialize their children successfully unless they have support from their extended community.[11] In such families, children need extensive contact with adults of the opposite sex if they are to be psychologically healthy.

Those in stepfamilies must recognize that these families are different from biological ones. Counselors often caution a stepparent not to attempt to discipline stepchildren, but to allow the biological parent do so. A child who has lost a parent through death or divorce is often resentful of her parent's new partner and will rebel or be distressed by that "outsider's" attempts to control or correct her. Some research has shown that it may take as many as eight years for a stepchild to bond with a stepparent.[12]

A third threat to the importance of the biological family is the neglect of family relationships and even the refusal to permit them. All too often grandparents are not allowed to spend time with their grandchildren, because their own children or their spouses do not value such contact. It is no surprise that in the Israeli commune (kibbutz) the requirement that all adults treat all children alike failed. Why? Grandparents could not resist the urge to coddle their own grandchildren.

Even when grandparents are alcoholics, abusers, or the like, it is not wise for parents to prevent their contact with grandchildren; parents may arrange supervised visits for the protection of the children. The same holds true for arranging contact for children with their biological parents in the case of divorce or separation.

One further threat to appreciating the importance of biological ties is the way so many deny the influence of their childhood family. Despite the evidence of research and experience,[13] many people refuse to see how their pres-

---

11. David Blankenhorn, *Fatherless America* (New York: Free Press, 1995).

12. Patricia Papernow, *Bonds without Blood: Stages of Development in Remarried Families* (New York: Gardner, 1987).

13. Selected sources are as follows: Children of divorce divorcing at higher rate: Richard J. Udry, *The Social Context of Marriage*, 3rd ed. (New York: Harper, 1974), p. 399. Abused

ent problems and traits have been shaped by their past. Some are misled by the inaccurate King James translation of 2 Corinthians 5:17, according to which at conversion "old things have passed away and all things have become new." Correctly, the passage concludes: "behold new things." Even after the new birth we carry some baggage from the past. A counseling colleague reports that he frequently encounters this denial in the older students he counsels. "Tell me about your childhood family," he asks, seeking to uncover some roots of the student's problem. "Oh, that doesn't matter," students typically reply, "I came to Christ in my early twenties; what happened before that is of no consequence."

It is of consequence. We must recognize and affirm that genetic relationships matter.

Affirming the Importance of Commitment to Marriage and Family

About four years into our marriage, my wife stunned me by disclosing that she really wanted to leave me during her first two years as my wife. I was surprised because I did not think those years were so bad. But she explained (and I can quote her precisely because shock etched her words in my memory): "I didn't leave you because I don't believe in divorce, God doesn't believe in divorce, and my Aunt Vea doesn't believe in divorce." These three reasons comprise the biblical concept of marital commitment. We are held together by a promise that is personal, divinely sanctioned, and sanctioned by others. My wife, Ginger, and I have something wonderful going today in our marriage. It developed, however, not because we were instantly compatible, but because we were intensely committed.

Proverbs 2:17 stresses two of these elements of the marriage covenant, maintaining that the adulteress has left the partner of her youth (the personal aspect) and ignored the covenant she made before God (the divine sanction).

---

becoming abusers: Byron Egeland, Deborah Jacobvitz, and Kathleen Papatola, "Intergenerational Continuity of Abuse," in Richard J. Gelles and Jane B. Lancaster, eds., *Child Abuse and Neglect: Biosocial Dimensions* (New York: Aldine de Gruyter, 1987), p. 270. Sons of alcoholics becoming alcoholics: L. Midanik, "Familial Alcoholism and Problem Drinking in National Drinking Practices Survey," *Addictive Behaviors* (1983): 133-41. Adults from dysfunctional families depressed: Sandra D. Wilson, "A Comparison of Evangelical Christian Adult Children of Alcoholics and Non-alcoholics on Selected Personality and Religious Variables." Doctoral Dissertation. Presented to the Union Graduate School of the Union for Experimenting Colleges and Universities, 1988.

Genesis 2:14, indicating a man will leave his father and mother to be united with his wife, teaches that the marital covenant is publicly announced. This is not to say that divorce is unforgivable or always unacceptable. Nevertheless, these Scriptures do affirm that commitment is the basis of marriage and is not to be taken lightly.

Yet, divorce statistics in the United States show that many are not taking marriage seriously. Not all the statistical evidence is bad news. Eighty percent of married people in the United States today still are married to their first mate.[14] Many still have a high regard for their marriage commitment. Yet, the laxity of dedication shows up when we compare couples today with those of the past. The percentage of divorce among people who married in 1880 was only 10 percent. However, those who married in 1950 have a divorce rate of 30 percent; those in 1970, 50 percent; and estimates of divorce among those who married in 1990 run as high as 67 percent.[15]

Not only is there weakening in the commitment to marriage, but there is some erosion of the commitment to other family members. Deadbeat dads (and they are numerous) not only fail to spend money to care for their offspring, but also fail to spend time relating to them. Although there is evidence that most adult children care for their aging parents, many neglect their commitment to them.

There are even trends in the evangelical church — long a bastion of marriage and family — that may foster the weakening of these commitments. For instance, there is a renewal of the discussion of the conflict between church and family. For years, we have argued whether the church or the home should have higher priority in a Christian's life. Pastors have often complained that people neglect the local church for their family activities. Now the controversy has been rekindled by Rodney Clapp's claim that the church should take a higher place in our lives. He contends:

> For years it has been popular among evangelicals to list three lifetime priorities, in this order: God, family and church. . . . In these popular rankings, family usurps the place that the New Testament assigns to the church.[16]

14. Andrew W. Greeley, *Faithful Attraction: Discovering Intimacy, Love, and Fidelity in American Marriage* (New York: Tom Doherty Associates, 1991), p. 18.

15. Daniel Goleman, *Emotional Intelligence* (New York: Bantam Books, 1995), p. 129; John Gottman, *What Predicts Divorce* (Hillsdale, N.J.: Lawrence Erlbaum Associates, Inc., 1993).

16. Rodney Clapp, *Families at the Crossroads* (Downers Grove, Ill.: InterVarsity Press, 1993).

Jesus has displaced the family, Clapp maintains, citing how on one occasion Jesus publicly refused to acknowledge his mother and siblings, insisting "Whoever does the will of my Father in heaven is my brother and sister and mother" (Matt. 12:48).

However, the Bible has much more to say on the matter. Other data from the New Testament do not suggest the local church should always be first. Paul, for instance, maintains that a widow should be cared for first by her own family, and that the church should care for her only when her family cannot (1 Tim. 5:14-16). Also, parents are responsible for the spiritual nurture of their children; it is not primarily the church's responsibility (Eph. 6:4).

A more biblical ordering of our priorities should place the kingdom of God first. There can be no argument against this, for Jesus explicitly commanded: "Seek first the kingdom of God" (Matt. 6:33). Jesus embraced this principle when he announced that those who do the will of God are his brothers, sister, and mother. However, the kingdom of God is not identical with the local church. The kingdom also includes the family. Therefore, a Christian who does the will of God is loyal to both local church and family commitments. Seeking first the kingdom of God can be done by going around the world as a missionary or it can include going around the block with one's child as a parent. It can include donating food to the church food bank or providing for one's own family.

As a general rule, we should not put either the family or the church first. Both are created by God and both contain privileges and responsibilities. At times the demands of one will clash with the demands of the other. Whether or not we give priority to one is determined by the circumstances. When there are pressing needs in the body of Christ, a person may sacrifice family for church. Yet, if an ill wife places heavy demands on a Christian's time and attention, his church involvement will be limited.

Just as Clapp's ordering of priorities is a threat to our commitment to marriage and family, his approach to caring for parents has a similar effect. He maintains that Christ has set adult children free from any responsibility for their parents. Because Christ has established a new order, the fifth commandment to honor parents by caring for them in their old age is apparently no longer in force. He maintains that Christians may choose to do so out of love, but they have no obligation. However, Jesus appeared to have no inclination in this direction. He castigated the Pharisees for ignoring their financial obligation to their parents by putting their estates in trust (Mark 7:9).

While his ideas may tend to weaken our family loyalties, Clapp does do us a service by his strong emphasis on the importance of loyalty to the local church. In fact, it is misleading to affirm the family in isolation, without affirming the larger context within which the family thrives.

Affirming the Importance of Community

Any affirmation of our commitment to family life should be placed side by side with an affirmation of our commitment to the church as a community. As many sociologists contend, the breakdown of families is not as great a problem as the breakdown of communities. Families, often in isolation and with little support, are being asked to do what normally was done by both family and community. It takes both a village *and* a family to raise children.[17]

For decades social commentaries have bemoaned our loss of community. As a result, Christians have been striving to turn their local churches into closely knit communities. They realize that we cannot be Christians alone. We need each other. In fact, advocates of community seem to agree, for the most part, on the kind of Christian communities we need to foster.

First, they should be intimate. Though the church in the strictest sense is not a family, it is to be family-like. We need to be involved in each other's lives and open to each other's hearts. People who lack the skills for family life can perhaps learn those skills through involvement in the life of the church. Moreover, there is an urgent need for intergenerational contact. Children need to be exposed to models of Christian behavior and to adults outside the family who love and influence them.

Second, Christian communities should be compassionate. As Dr. David Wells has observed, the best apologetic in postmodern times is to offer Christ as the answer to the problems of broken, distraught people.[18] This approach is not new. "Come to me, all you who are weary and burdened," said Jesus "and I will give you rest" (Matt. 11:28). As Joseph Parker once put it so well: "Preach to the suffering and you'll never lack a congregation."[19] Churches

17. Books in addition to Clapp's that stress community life as necessary for family include David Mace and Vera Mace, *The Sacred Fire: Christian Marriage through the Ages* (Nashville: Abingdon Press, 1986); Mark DeVries, *Family-Based Youth Ministry* (Downers Grove, Ill.: InterVarsity Press, 1994); Cameron Lee, *Beyond Family Values: A Call to Christian Virtue* (Downers Grove, Ill.: InterVarsity Press, 1998).

18. David Wells, "The Crisis of Truth" and "What Is Right? Our Public Predicament and God's Holiness." Audio and video tapes available from Bannockburn Institute, 2065 Half Day Road, Bannockburn, IL 60015.

19. Quoted in David B. Biebel, *Jonathan You Left Too Soon* (Nashville: Thomas Nelson, 1981), p. 9.

should display the kind of sympathy, understanding, and caregiving that makes hurting people feel welcome.

Much is at stake if society continues to allow all uses of reproductive technologies without regard for their impact on the family. We do need to oppose certain technologies. Ultimately, however, our greatest impact may well come from not what we stand against, but what we stand and strive for: strong genetic ties, firm commitment to marriage and family, and the building of intimate, compassionate communities.

Index